PLACE IN RETURN BOX to remove this checkout from your record.
TO AVOID FINES return on or before date due.

DATE DUE	DATE DUE	DATE DUE

Directing
Ecological
Succession

Directing Ecological Succession

JAMES O. LUKEN
Department of Biological Sciences,
Northern Kentucky University,
USA

CHAPMAN AND HALL
LONDON • NEW YORK • TOKYO • MELBOURNE • MADRAS

UK	Chapman and Hall, 2–6 Boundary Row, London SE1 8HN
USA	Chapman and Hall, 29 West 35th Street, New York NY10001
JAPAN	Chapman and Hall Japan, Thomson Publishing Japan, Hirakawacho Nemoto Building, 7F, 1-7-11 Hirakawa-cho, Chiyoda-ku, Tokyo 102
AUSTRALIA	Chapman and Hall Australia, Thomas Nelson Australia, 480 La Trobe Street, PO Box 4725, Melbourne 3000
INDIA	Chapman and Hall India, R. Seshadri, 32 Second Main Road, CIT East, Madras 600 035

First edition 1990

© 1990 James O. Luken

Typeset in 11/12½pt Sabon by EJS Chemical Composition, Bath, Avon
Printed in Great Britain by The University Press, Cambridge

ISBN 0 412 34450 5

All rights reserved. No part of this publication may be reproduced or transmitted, in any form or by any means, electronic, mechanical, photocopying, recording or otherwise, or stored in any retrieval system of any nature, without the written permission of the copyright holder and the publisher, application for which shall be made to the publisher.

British Library Cataloguing in Publication Data

Luken, James, O.
  Directing ecological succession.
  1. Ecological succession
  I. Title
  574.5

  ISBN 0–412–34450–5

Library of Congress Cataloguing-in-Publication Data

Luken, James O.
  Directing ecological succession/James O. Luken.—1st ed.
    p. cm.
  Includes bibliographical references.
  ISBN 0–412–34450–5
  1. Plant succession. 2. Vegetation dynamics.
  3. Ecological succession. I. Title
QK910.L85   1990
581.5′247—dc20   90–1781
                                                                              CIP

Contents

Foreword		ix
1	Succession management: an introduction	1
	1.1 Introduction	1
	1.2 What do we now know about succession?	2
	1.3 A history of early succession management	6
	1.4 Modern succession management	9
	Summary	15
2	Obtaining information on succession	19
	2.1 Introduction	19
	2.2 Published research	20
	2.3 Cooperative research agreements	20
	2.4 On-site demonstrated research	22
	2.5 Methods to study succession	23
	Summary	32
3	Plant populations: growth, decline, and persistence during succession	35
	3.1 Introduction	35
	3.2 Population trends during succession	35
	3.3 Population characteristics and measurements	36
	3.4 Population growth and persistence from seeds	40
	3.5 Population change from vegetative reproduction	49
	3.6 Population decline by death	54
	Summary	56
4	Methods of managing succession: plant and plant part removal	59
	4.1 Introduction	59
	4.2 Mowing, clipping, and cutting	60
	4.3 Herbicides	73
	4.4 Fire	84
	4.5 Cabling	97
	Summary	100

5 Methods of managing succession: changing resource availability — 103
5.1 Introduction — 103
5.2 Soil nutrients — 103
5.3 Nutrient exhaustion — 111
5.4 Water — 114
Summary — 125

6 Methods of managing succession: changing propagule availability — 127
6.1 Introduction — 127
6.2 Seeding and mulching — 128
6.3 Topsoiling — 141
6.4 Unmanaged succession following severe disturbance — 146
Summary — 148

7 Animals and succession — 151
7.1 Introduction — 151
7.2 Animal communities: changes during succession — 151
7.3 Ecological succession: changes due to animal activities — 163
7.4 Seed dispersal by animals — 176
Summary — 177

8 A landscape perspective — 179
8.1 Introduction — 179
8.2 Succession management in nature reserves — 179
8.3 The internal dynamic landscape of a nature reserve — 180
8.4 Optimum management unit patch size — 183
8.5 Distribution and connection — 189
8.6 External factors — 192
Summary — 198

9 Information systems for prediction and decision-making — 201
9.1 Introduction — 201
9.2 Succession models — 201
9.3 Expert systems — 209
9.4 Resource management algorithms — 211
Summary — 215

Afterword	217
References	219
Index	245

Foreword

This book is a manual for the practicing natural resource manager who commonly works with plant communities. It begins by developing a general model of succession management that is applicable to any ecosystem that supports plant life (Chapter 1). Implementation of this model relies on the establishment of working relationships between ecologists and resource managers, a subject treated in Chapter 2. In addition, anyone that attempts to manage succession must have at least a basic understanding of plant population biology. As such, in Chapter 3, I attempted to provide a few population biology concepts that are relevant to succession management. Chapters 4–7 deal with specific methods of managing succession as well as the effects of these management activities on succession. Examples are taken from plant communities around the world, but the reader may note a certain bias towards North American plant communities. This should not suggest that we in the US are at the forefront of informed succession management. Rather, it simply reflects the types and amounts of information that were available to me. The importance of scale in resource management problems is treated in Chapter 8 by taking a landscape perspective. I hope that attempts to manage succession in the future do not overlook this important and rapidly developing field. Lastly, Chapter 9 examines some information systems that can be used to make decisions about succession management.

The impetus for this book comes from the discrepancy that commonly exists between the goals of practicing ecologists and the goals of practicing resource managers. The former group generates tremendous amounts of information regarding succession—information that could be of great value to the latter group. Often, however, the transfer of information does not occur because people from the two groups do not interact, they read different journals, and they attend different conferences. This book is an attempt to build both concept and information linkages between ecologists and resource managers.

I have attempted to summarize and make generalizations where and when the existing data allow it. Often, however, data are contradictory and one must accept the fact that the response of a particular system to management may not be predictable. This in itself speaks loudly for the necessity of site-specific research before large-scale management programs are implemented. I hope ecologists in the future participate fully in such research.

Several reviewers provided me with constructive criticisms of the entire manuscript. I thank Reed Noss, John Thieret and Gary Wade for these efforts that greatly improved the final product. Responsibility for subsequent omissions or mistakes rests squarely on my shoulders.

Other individuals unselfishly provided me with photographs and reprints. For these materials I thank Clark Ashby, Donald Beck, W.R. Byrnes, Walter Carson, William Connor, Edward DePuit, Gerritt Heil, Paul Kalisz, John Leigh, I.A.W. Macdonald, Robert Marrs, Henry Murkin, William Niering, Christer Nilsson, M.J.M. Oomes, Edward Redente, M.B. Usher, Gary Wade, Carl Wambolt, Neil West, Dick Williams, Robert Wittwer, M. Karl Wood and Willis Vogel.

Numerous publishers and authors granted permission to reprint copyrighted materials and I thank them for facilitating this transfer of information.

I am grateful to Beth Merten and her staff for the countless hours spent reproducing figures. I am also indebted to Sharon Taylor who efficiently obtained hundreds of articles and books through interlibrary loan.

Northern Kentucky University supported the writing of this book through a summer fellowship and the Department of Biological Sciences provided support for manuscript preparation.

Lastly, I must thank my wife Linda who patiently read the entire manuscript and provided me with valuable editorial comments. She also encouraged me at times when the problems appeared insurmountable.

James O. Luken

1

Succession management: an introduction

1.1 INTRODUCTION

Ecologists have long noted and studied the tendency of plant communities to change through time. This process, known as succession, was first reviewed by Clements (1916) who provided an historical description of succession research dating back to 1685. Since the publication of Clements' review, succession continues to be a fertile area for the development of hypotheses, models, generalizations, and predictions.

An incisive paper tracing the development and senescence of succession theory (McIntosh, 1980) began with the following statement:

> Succession is one of the oldest, most basic, yet still in some ways, most confounded of ecological concepts. Since its formalization as the premier ecological theory by H.C. Cowles and F.E. Clements in the early 1900s, thousands of descriptions of, commentaries about and interpretations of succession have been published and extended inconclusive controversy has been generated.

This quote, when juxtaposed against an abundant published literature on succession, should not foster pessimism about the work of ecologists who study succession. Rather, it should galvanize the fact that change in plant communities through time is a complex process far from being totally understood or predicted.

Although much is now known about succession, the application of this knowledge in resource management has not progressed rapidly. Part of the blame for this can be traced to ecologists themselves who do not wish to work in managed systems where the results of their research are site-specific and may not readily lead to new ecological theories (Smith, 1988). The remainder of the blame, however, must be shared among ecologists and resource managers who fail to cooperate in the solution of natural resource management problems involving

2 An introduction

succession. Indeed, any such problem that involves plant communities also involves succession.

Resource managers require information that can be applied to their specific situations. Ecologists on the other hand, are interested in the 'whys?' or 'wheres?' of succession so that new explanations can be generated. In the arena of determining a management plan, theoretical considerations are secondary to simply stated management goals and recommendations. This book is an attempt to fill the void existing between the scientific interpretation of succession and the application of this knowledge in resource management.

There is now nearly a century of research where successional patterns are described from numerous plant communities scattered around the earth. Palaeoecology is providing new insights into the historical dynamism of plant communities and the effects of prehistoric people on such changes. The role of natural disturbance in plant-community dynamics is now well established (White and Pickett, 1985). An abundant literature, especially from western Europe and western US, details many techniques used to control succession. Computer models exist that successfully predict forest succession trends. And last, more mechanistic explanations of succession based on life histories of individual species are emerging to enhance the understanding of why plant communities change. This information will be used to provide the state and potential of succession management. Hopefully, the ideas presented here will demonstrate the futility of plant-community management over the long term without first including succession as an integral part of a management plan.

1.2 WHAT DO WE NOW KNOW ABOUT SUCCESSION?

Plant communities in all parts of the world change as they age. These changes can involve species replacements, shifts in population structure, and changes in availability of resources such as light and soil nutrients. Beyond the simple observation that plant communities change through time, however, is a staggering variety of explanations, all of which attempt to answer the following question: Why does succession occur?

One of the first descriptions of succession as a sequence of identifiable processes was presented by Clements (1916). He proposed that six basic processes function during succession: nudation, migration, ecesis, competition, reaction, and stabilization (Figure 1.1). Nudation, the process creating a patch of bare soil, is what begins succession. Plants colonizing disturbed patches will either come from propagules (i.e., seeds, root fragments, or whole plants) remaining in the soil or propagules migrating from somewhere else. Ecesis, the successful establishment of plants, is controlled by local environmental conditions and the characteristics of plant species available at the site. Competition, the battle for limiting resources that occurs among established plants, will eliminate some species and favour others during establishment or during later stages of succession. Reaction, the change of the environment as a result of plants growing and dying, will continually change the availability of resources. Stabilization, a condition that rarely if ever occurs, develops as very long-lived species dominate a site. The approach of Clements – that is, separation of succession into component processes or events – has been taken by different researchers to explain succession under a variety of environmental conditions.

The traditional interpretation of succession that developed after Clements' work and a view that has been much maligned – and properly so – is that of 'relay floristics'. One group of plant species establishes and is then replaced by another group until a stable state is achieved (Figure 1.1). Although this interpretation was long the standard paradigm for teachers describing succession to students, it is mostly a teaching convenience and does not adequately describe the process.

Egler (1954) stressed the importance of the propagule pool in controlling succession. This explanation of succession termed 'initial floristics' was formally developed after he studied secondary succession in old fields. He proposed that when vegetation is disturbed, the soil retains a propagule pool representing previous successional stages (Figure 1.1). Although plant species from previous successional stages may be present as seedlings soon after the initial disturbance, the different stages of a successional pathway are simply different species groups assuming dominance through time. The absence of a species in the original propagule pool means that this species will not be a part of succession or that it will become a part only very slowly. Thus, if the propagule or species pool early in succession can be successfully manipulated, then indeed, the mix of plant species participating in succession can be modified.

Drury and Nisbet (1973) viewed succession as a process in which plant species are sorted along a gradient of resources (Figure 1.1). Because each individual species has a unique optimum for growth or reproduction and because resource (e.g., light and nutrients) availability changes through time, species replacement occurs. Pickett (1976) expanded this argument by incorporating competition into the time/resource gradient originally proposed by Drury and Nisbet

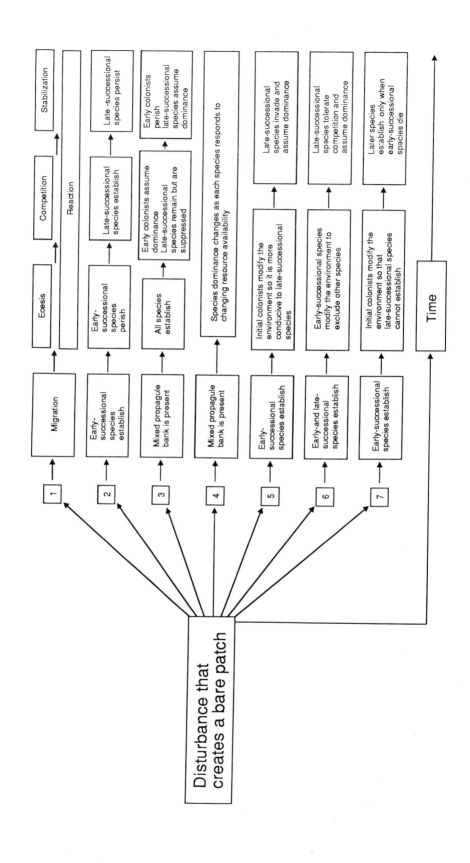

(1973). Thus, species replacements occur during succession in accord with the outcome of competition in a changing environment. Pickett's interpretation was later supported by a 20-year record of succession in permanent plots (Pickett, 1982).

Clearly, one of the most influential papers attempting to develop explanations for succession was written by Connell and Slatyer (1977). They developed three models for succession – 'facilitation', 'tolerance', and 'inhibition' – and then reviewed the literature to determine the amount of evidence in support of each model. 'Facilitation', the traditional interpretation accepted by many researchers at the time Connell and Slatyer did their work, is succession in which plant species prepare the way for other plant species (Figure 1.1). Surprisingly, it appears to operate only in those situations where one plant species greatly improves the nutrient status in the soil, e.g., when nitrogen-fixing plants colonize nutrient-poor glacial till or river sand. The 'tolerance' model proposes that plant species entering successional pathways after the initial colonists are established must be capable of surviving under low resource availability. Late successional species are those that can tolerate low resource availability and eventually emerge into a dominant position. There were few real-world situations that appeared to fit this model. Last, 'inhibition' is where established species inhibit the invasion and growth of all potential competitors during the entire duration of a successional pathway. There was much evidence in support of this model. Many failed attempts to speed the invasion of trees into established vegetation will attest to the fact that inhibition is a reality.

It is likely that still more generalized explanations for succession will emerge in the future; the approach, however, will be different. Instead of searching for models with universal utility, researchers will focus more and more on the individual species participating in succession. This trend is already in evidence. Noble and Slatyer (1980) attempted to define 'vital attributes' of species that would predict their performance during succession. These attributes included the method of arrival or persistence after a disturbance, the ability to enter an existing community and then grow to maturity, and the time it takes for a species to attain critical stages in its life cycle. Pickett *et al.* (1987)

Figure 1.1 Various interpretations of succession. 1. The Clementsian view. 2. Relay floristics. 3. Initial floristics (Egler, 1954). 4. Changing resource availability (Drury and Nisbet, 1983). 5. Facilitation (Connell and Slatyer, 1977). 6. Tolerance (Connell and Slatyer, 1977). 7. Inhibition (Connell and Slatyer, 1977).

Table 1.1 General causes of succession, contributing processes, and modifying factors

General causes	Contributing processes	Modifying factors
Site availability	Disturbance	Size, severity, time, dispersion
Species availability	Dispersal	Landscape configuration, dispersal agents
	Propagules	Land use, Time since last disturbance
Species performance	Resources	Soil, topography, site history
	Ecophysiology	Germination requirements, Assimilation rates, Growth rates, Genetic differentiation
	Life history	Allocation, Reproductive timing and mode,
	Stress	Climate, site history, Prior occupants
	Competition	Competition, herbivory, resource availability
	Allelopathy	Soil chemistry, microbes, Neighbouring species
	Herbivory	Climate, predators, Plant defences and vigour, Community patchiness

From Pickett *et al.* (1987). Used by permission.

expanded on the concept of vital attributes and developed a hierarchy of succession including the causes of succession, the contributing processes, and the defining factors (Table 1.1). This hierarchy is solidly based at the level of individual species and the factors controlling their availability and performance on a site.

1.3 A HISTORY OF EARLY SUCCESSION MANAGEMENT

Before developing a generalized approach to succession management it is informative to examine how people have historically managed succession in pre-modern times. Indeed, the word 'manage' is used incorrectly here. As will be obvious, succession was changed or

A history of early succession management

modified largely as a result of disturbances created by people in their quest for food, fibre or fuel. Succession was 'managed' only fortuitously. Two examples will be given to demonstrate how humans have historically modified succession using relatively simple technologies.

1.3.1 Anthropogenic influence: southern Scandinavia

Berglund (1969) presented a summary of palaeoecological data collected in southern Scandinavia and detected four major periods of expanding human influence on plant communities there during the last 5000 years. An early Neolithic landnam period (5300–4500 years Before Present (BP)) was characterized by decreases in tree species such as basswood, elm, and ash with concomitant increases in pasture and tilled land. Deforestation and the conversion of forests into pasture was indicated by the sudden appearance of pollen from herbs such as *Plantago lanceolata*, *Rumex acetosella*, and various grasses. Conversion of forest into tilled land was indicated by the pollen of herbs such as *Centaurea cyanus* and *Plantago major*. Succession after clearing produced regenerating woodlands with few elms but increased presence of birch and beech. The middle Neolithic (4200–3000 years BP) saw further expansion of agriculture and grazing. The raising of livestock in particular caused the retrogression of many forests, which were converted into shrub-dominated or heath communities. In early and late Iron Age periods (2200–1500 years BP and 1200–900 years BP, respectively) there was widespread conversion of forest to more open heathland, *Juniperus*-heath, or birch-dominated shrubland and also the expansion of cultivated land.

Berglund concluded that agricultural land and pastures were periodically 'laid waste' during times when human influence declined and the land was presumably abandoned. Thus for nearly 5000 years people in Scandinavia set back and then released succession. The net result was apparently a change in the pattern and direction of plant community development although a changing climate and the migration of species may also have played a role (Berglund, 1969).

1.3.2 Anthropogenic influence: eastern deciduous forest, USA

Chapman *et al.* (1982) documented the prehistoric impact of American Indians on the successional status of plant communities near the Little Tennessee River in the United States. Coinciding with the appearance of cultivated crops such as squash (4400 years BP) and gourds (3300 years BP) was an increase in disturbance-favoured plant species such as cane (*Arundinaria*) and yellow poplar (*Liriodendron tulipifera*). Bottomland trees such as maple and birch declined in importance.

Indians apparently began clearing the nearly continuous forests on lower river terraces to grow crops. Subsequent periods of abandonment created vegetation patches in various stages of succession. Closed forests on higher terraces remained relatively undisturbed. The opening of forests near the river also created more early successional communities and more edge habitat that may have increased the carrying capacity of the land for white-tailed deer. Chapman *et al.* (1982) hypothesized that the growing of crops and the greater availability of deer facilitated the expansion of the Indian population during this time.

1.3.3 Anthropogenic influence: generalizations

Emerging from pollen, artifact, and macrofossil evidence collected around the world is the conclusion that people, where they congregated, were a strong force in vegetation change during the last 5000 years. This was a period of time when anthropogenic disturbance exceeded the rate of natural disturbance in many parts of the world. Succession, once operating as a repair process following natural disturbance, was pushed to new limits in the face of landscape conversion. With this long documented history of human disturbance, it is likely that modern successional pathways we now observe are expressing the prehistoric as well as the historic influences of people. Some generalizations about these prehistoric influences can be made.

1. Deforestation was the primary effect of human activity. Early deforestation was no doubt linked to the use of wood for fuel and as a raw material. Later, trees were removed to create pastures and agricultural land.
2. Periods of intense human activity were associated with the appearance and disappearance of various plant species. People drove certain species to near extinction while at the same time facilitating range extensions of other species (Delcourt, 1987).
3. Often, man-made clearings were periodically abandoned. Thus landscapes, at least in forested areas, were changed from nearly closed forests to a mosaic of patches in various stages of succession. Disturbance–favoured taxa became important in these early successional patches (Delcourt, 1987).
4. Human impact on plant communities intensified until historical times and has continued to this day.

1.4 MODERN SUCCESSION MANAGEMENT

Presently, many resource management problems exist where succession can, and should, be an integral part of the planning strategy (Table 1.2). Incorporating succession management in natural resource management decision-making is the next step after careful description of successional pathways (Westhoff, 1971; Slatyer, 1977; White and Bratton, 1980). Unfortunately, few writers have provided generalized instructions for how this should be accomplished (Slatyer, 1977; Rosenberg and Freedman, 1984)). Moreover, there are complex questions regarding the distinction between 'vegetation management' and 'succession management'.

Although modifications of plant communities by prehistoric people were deliberate and widespread, corresponding changes in successional pathways and subsequent landscape modifications were largely unplanned. This can be said of many situations in present times when plant communities are managed to achieve short-term goals

Table 1.2 Some resource management problems where succession can be manipulated to achieve management goals

---

Conserving rare or endangered species.
Conserving and restoring communities.
Manipulating the diversity of plant and animal communities.
Creating relatively stable plant communities on rights-of-way.
Revegetating drastically disturbed lands.
Minimizing the impact and spread of introduced species.
Maximizing wood production from forests.
Minimizing adverse environmental effects of forestry.
Predicting fuel buildup and fire hazards.
Increasing animal populations for recreation and aesthetics.
Developing multiple use plans for parks and nature reserves.
Determining the minimum size of nature reserves.
Minimizing the impact of roads, parking areas, trails and campsites.
Preserving scenic vistas in parks.
Minimizing the cost of grounds maintenance on public lands.
Controlling water pollution.
Minimizing erosion.
Maintaining high quality forage production.
Minimizing the cost of crop production in agricultural communities.
Developing vegetation in wetlands or on the edges of reservoirs.

---

10 An introduction

without consideration of succession (i.e., vegetation management). For example, thousands of hectares of land each year are blanket-sprayed with herbicides to eliminate existing woody vegetation. This management strategy is short-sighted and does not consider that plant communities may recover in a new and perhaps more troublesome state.

Much information exists on the effects of humans on plant community composition and these influences fall under the general heading of succession management if the following principles are understood.

1. All plant communities show some form of succession at all times, i.e. there are no stable plant communities (Niering, 1987).
2. Management activities modify the rate and direction of succession rather than the state of vegetation.

An approach for managing plant communities is needed that will draw on the expanding succession database. Ideally, a succession management model should be able to accommodate most resource management situations; it should be sensitive to the diversity of current interpretations of the succession process (Drury and Nisbet, 1973; Connell and Slatyer, 1977; Noble and Slatyer, 1980; Peet and Christensen, 1980; Walker and Chapin, 1987; Pickett *et al.*, 1987); and it should be sufficiently general to include the management of natural factors affecting succession and of human-directed factors.

1.4.1 A three-component succession management model

In an attempt to build a succession management model applicable to a wide range of resource management situations, Rosenberg and Freedman (1984) proposed five sequential steps: designed disturbance, selective colonization, inhibitory persistence, removal, and regeneration. It is clear that Rosenberg and Freedman were strongly influenced by Clements (1916) as well as by Connell and Slatyer's (1977) inhibition model. Their attempt to link the causes of succession with methods of controlling succession is commendable. However, a simpler approach can be taken that has wider applicability in resource management.

Pickett *et al.* (1987) maintained that there are three basic causes of succession: site availability, differential species availability, and differential species performance (Table 1.1). These three components contributing to natural succession can be modified for application in resource management situations. To manage succession, three

Table 1.3 Three components of succession management corresponding to three general causes of succession

Three components of succession management	Three general causes of succession
1. Designed disturbance	1. Site availability
2. Controlled colonization	2. Differential species availability
3. Controlled species performance	3. Differential species performance

Modified from Pickett *et al.* (1987). Used by permission.

components are required: designed disturbance, controlled colonization, and controlled species performance (Table 1.3). Designed disturbance includes activities initiated to create or eliminate site availability. Controlled colonization includes methods used to decrease or enhance availability and establishment of specific plant species. Controlled species performance includes methods used to decrease or enhance growth and reproduction of specific plant species.

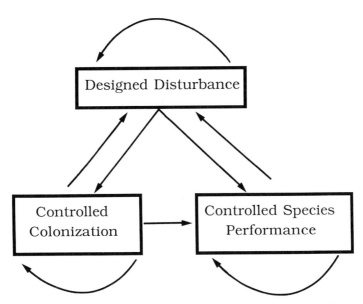

Figure 1.2 Three components of a succession management model. Straight lines indicate sequential steps; curved lines indicate repeated steps. Modified from Rosenberg and Freedman (1984).

12 An introduction

The components of this model should not be construed as a stepwise process because not all resource management situations initially require a designed disturbance. But succession management is an ongoing process. Thus, there must be movement from one component to the other, or there may simply be repeated applications of a single component through time (Figure 1.2).

1.4.2 Designed disturbance

Designed disturbance is derived from the observation that disturbance of some sort usually initiates successional pathways. Clements (1916) referred to this as 'nudation'. Although anthropogenic clearing of vegetation and detritus to expose bare ground was common during the Neolithic period and is still common today, there are numerous natural disturbances that initiate or set back succession (Table 1.4). Indeed, the fact that people accelerated the rate of disturbance has

Table 1.4 Natural disturbances that commonly initiate or modify successional pathways in plant communities

Disturbance	Ecosystem or geographic area
Fire*	Boreal, temperate, and western montane forests, grasslands, chaparral, coastal plain, heath, fynbos shrubland, Mediterranean shrubland
Wind storms	Temperate forest
Ice storms	Temperate forest
Cryogenesis	Arctic and alpine tundra
Thermokarst*	Arctic tundra, boreal forest
Flash floods	Desert
Alluviation*	River valleys
Landslides*	Areas of steep topography and unstable substrates
Extreme climatic conditions	All areas
Grazing*	Grassland, heath, savannas, shrublands, tropical, temperate and boreal forests
Pathogens*	Heath, grassland, tropical, temperate, boreal and montane forests

* Natural disturbances that are commonly eliminated or accelerated by the activities of people.
Modified from White and Pickett (1985). Used by permission.

probably made succession and successional plant communities more common than if only natural disturbances were operating. The word 'natural' here refers to processes not commonly controlled by people, but in many instances the intensity and frequency of even these natural disturbances have been changed. A good example is wildfire and its modification by fire suppression policies. Natural disturbances as well as the direct activities of people will determine the availability of sites where succession can occur (Pickett et al., 1987).

White and Pickett (1985) defined disturbance broadly as 'any relatively discrete event in time that disrupts ecosystem, community, or population structure and changes resource, substrate availability, or the physical environment.' Rosenberg and Freedman (1984) did not formally define 'designed disturbance', but it can be considered any disturbance (sensu White and Pickett, 1985) produced or manipulated by people to initiate and subsequently control a succession pathway. As is true for natural disturbance (White and Pickett, 1985) the effect of a designed disturbance will depend on severity, size, timing and the type of disturbance that is used.

Designed disturbances can take many forms. Those listed in Table 1.5 include person-generated disturbances that have been used to initiate succession, but designed disturbance can also result when

Table 1.5 Various person-generated designed disturbances that have been used to initiate succession

Designed disturbance	Reference
Burning	Mallik and Gimingham (1983)
Bulldozing	Schott and Pieper (1987)
Cabling	Rippel et al. (1983)
Chopping	Moore et al. (1982)
Clipping	Marrs (1987a)
Flooding and draining	Franz and Bazzaz (1977)
Herbicide application	Sterrett and Adams (1977)
Ploughing and rotovation	Marrs (1987a)
Scraping	Jefferson and Usher (1987)
Scraping and topsoil mixing	Doerr et al. (1984)
Slash and burn	Uhl (1987)
Soil compaction	Prose et al. (1987)
Temporally staggered tillage	Altieri (1981)
Thermal shock	Dunn and Sharitz (1987)

14 An introduction

people intentionally modify natural disturbances such as fire, flooding, and pathogens in order to change succession. Generally, natural disturbances are modified to eliminate the disturbance rather than to change succession. The utility of any specific designed disturbance in succession management will depend on the plant community type, the site history, the season, the macro- and microclimate, the local successional pathway, and the management goals. Clearly, management goals must be established before a designed disturbance is selected and implemented.

1.4.3 Controlled colonization

Since controlled colonization is the manipulation of plant species availability and establishment, it is not surprising that many of the methods used to control colonization may also be used as designed disturbances (Tables 1.5 and 1.6). Methods for controlled colonization that do not qualify as designed disturbances include techniques augmenting the propagule and species pools such as direct planting (Figure 1.3), seeding, and topsoiling and those techniques, such as the use of soil fabrics, directly inhibiting colonization of some species. As with designed disturbance, controlled colonization also

Table 1.6 Various techniques that have been used to control colonization during succession management

Management technique	Reference
Burning	Marrs (1987b)
Broadcast seeding	Doerr and Redente (1983)
Cutting	Marrs (1987b)
Direct planting	Hatton and West (1987)
Drill seeding	Doerr and Redente (1983)
Fertilization	Persson (1981)
Grazing	Gibson et al. (1987)
Herbicide spraying	Weber et al. (1974)
Irrigation	Persson (1981)
Rotovating	Marrs (1987b)
Scraping	Jefferson and Usher (1987)
Soil fabrics	Grime et al. (1971)
Topsoiling	Hatton and West (1987)
Water level change	Smith and Kadlec (1983)

results when natural processes affecting species availability and establishment are intentionally modified to change succession. For example, planting shrubs to attract seed-carrying animals. The potential power of controlled colonization in succession management must be cast in light of two factors: the propagule pool and the 'initial floristics' of a site.

The availability of species at a site is ultimately controlled by the propagules present on the site or by propagules that can quickly arrive from other sites. The propagule pool consists of seeds, spores, rootstocks, cut stumps, bulbs, rhizomes, plant fragments, and even entire plants. Depending on the species, each of these can potentially give rise to a new individual. In controlled colonization, management activities must be aimed at the propagules, at the factors that move propagules (i.e., dispersal agents), or at the establishment of seedlings from the propagules.

1.4.4 Controlled species performance

After plant species are established and colonization is complete, artificial control of species performance is the appropriate component of succession management. Processes or conditions giving rise to differential species performance include physiological characteristics of the species, the life histories of the species, intra- and interspecific competition, allelopathy, herbivory, predation and pathogens (Pickett *et al.*, 1987). Along the same line, controlled species performance during succession management includes person-generated activities that modify the growth and reproduction of plants in a successional pathway (Table 1.7).

SUMMARY

During the last 5000 years, interactions between succession and people have taken many forms. In prehistoric times, people managed vegetation through cutting, burning, tilling and the raising of stock. They also inadvertently managed succession. As the activities of people became the dominant disturbance in many areas of the world, the successional character of landscapes was changed. With the expansion and contraction of human influence, landscapes became a mosaic of patches in various stages of succession. This trend continued to historic times where at present most plant communities, except those isolated or saved from human influence, are in some stage of recovery following human disturbance.

16 An introduction

Figure 1.3 Succession on Kentucky minespoil after surface grading and planting of mixed hardwoods and pines. The most successful tree species at this site was yellow poplar (*Liriodendron tulipifera*) with medium to good site indices. (a) The site after grading. (b) After 6 years of succession. (c) After 18 years of succession. Photographs courtesy of the US Forest Service and Dr Willis Vogel.

Succession has been studied intensively during the last 100 years. Successional pathways are described from plant communities around the world, and a number of theories exist to explain why succession occurs. The fact that plant communities are not static, however, creates resource management problems, the solution of which depends on (1) a basic understanding of the successional pathway involved and (2) information about how plant species in the successional pathway respond to different management techniques. Controlling the succession process to meet management goals is known as succession management.

A succession management model is presented that includes three components: designed disturbance, controlled colonization, and controlled species performance. Designed disturbance includes techniques used to create or eliminate sites where succession can be initiated. Controlled colonization includes techniques used to increase or decrease the availability and establishment of plant species.

Table 1.7 Various techniques that have been used to control species performance during succession management

Management technique	Reference
Burning	Mallik and Gimingham (1983)
Cabling	Rippel et al. (1983)
Excluding grazers	Davis et al. (1985)
Fertilization	Heil and Diemont (1983)
Grazing	Gibson et al. (1987)
Herbicide application	Marrs (1987a)
Irrigation	Doerr and Redente (1983)
Mowing	Vestergaard (1985)
Reducing soil fertility	Marrs (1985b)
Selective cutting	Lowday (1987)
Water level change	Bakker et al. (1987)

Controlled species performance includes techniques used to increase or decrease the growth and reproduction of plant species. The three components of this model can be adapted to a wide variety of resource management situations where succession must be modified.

2

Obtaining information on succession

2.1 INTRODUCTION

Much of what is now known about succession can be traced to descriptive and experimental research on a few types of plant communities (e.g., old fields, sand dunes or glacial moraines). Before succession management moves from the pages of journals or books to actual practice, more research on succession in anthropic (human-influenced) communities must be done. Such research requires close interaction between ecologists and resource managers in order to obtain the proper kinds of information.

A broad range of information is required to successfully implement a succession management programme (Slatyer, 1977). Some of the information can be described as 'baseline data'. For example, descriptions of successional pathways following various natural disturbances are ncessary before succession in these pathways can be manipulated. Other types of information include results of applied research where the effects of specific management techniques are tested. Non-scientific information regarding management goals, management costs, available resources, recreational uses of the land, public opinion, and existing laws is also necessary.

Resource management algorithms and expert systems (described in Chapter 9) provide systematic processes for decision-making that consider many different types of information including scientific, sociological, and economic information. Unfortunately, it is lack of scientific information that commonly hinders the success of such decision-making processes. This chapter will focus on techniques used by resource managers to obtain basic and applied scientific information on succession that is site- and problem-specific.

Three options are available to accumulate a database on succession in a particular area or plant community: locating the information in published literature, initiating research through contractual or cooperative agreements with ecologists, and self-initiated, on-site research.

2.2 PUBLISHED RESEARCH

The simplest and least-expensive method of obtaining information about local successional pathways is by locating published studies. Descriptive studies will simply detail or explain potential plant community changes occurring at a site or in a general area. Some descriptive studies date back to the early 1900s and must be used with caution because changing disturbance regimes and the invasion of introduced plant species can modify successional pathways. Therefore, the more recent a study, the more likely it is to reflect accurately the actual successional pathways.

Applied research is site-specific and examines the effects of management techniques on plant community development. Most studies of this type present data collected over a period of 2–5 years; the best of such studies track community development after treatment initiation for longer periods of time. Seldom does a single study include both descriptive and applied research.

With both descriptive and applied succession research there will always be questions regarding the applicability or relevance of data collected in one area and then applied to another area. As an example, consider the role of sumach (*Rhus*) species in facilitating the invasion of late successional tree species during grassland and old-field succession in the US. Two studies, one carried out in Oklahoma (Petranka and McPherson, 1979) and the other in Michigan (Werner and Harbeck, 1982) concluded that *Rhus* species facilitated the invasion of trees by eliminating competition from grasses and by attracting seed-carrying animals. From this information, one might conclude that tree invasion could be accelerated by establishing *Rhus* species. However, when colonies of *Rhus* growing on revegetated roadside embankments in Kentucky were examined for seeds, seedlings, or saplings of late successional trees, few were found (Luken and Thieret, 1987). The background of introduced species, the disturbance regimen, and the landscape position of the roadside *Rhus* colonies apparently changed the successional role of this species. Clearly, off-site descriptions of succession in the literature can provide useful information but the most relevant published data are those collected on-site.

2.3 COOPERATIVE RESEARCH AGREEMENTS

If no published data can be located on the successional problem under consideration, the task of collecting either basic or applied data can be given to trained ecologists. Often, local colleges and universities employ a number of individuals with expertise in ecology who are willing to carry out research but do not because they are unaware of the research needs or management problems of local resource managers. Slatyer (1977) presented a scheme outlining the potential role of ecologists in providing advice and expertise to resource managers when building of predictive models and when developing succession management plans. He pointed out that communication channels need to be opened between ecologists and resource managers. In addition, the individuals providing advice, funding, or labour to ecologists and individuals who constrain the activities of the resource manager (Figure 2.1) need to be brought into the decision-making process.

The level of expertise required to carry out research varies with the research goal. Simple descriptive studies of successional pathways can be handled by undergraduates who have taken general ecology, plant ecology, or ecological methods classes. Such studies also require

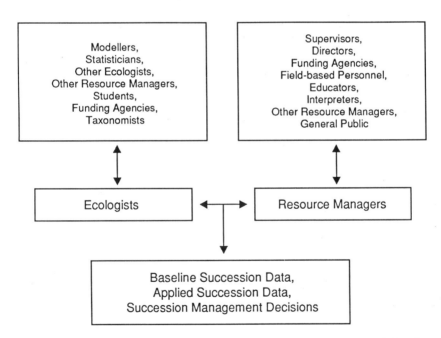

Figure 2.1 Communication pathways required to collect data and develop succession management decisions. Boxes at the top indicate individuals interacting with ecologists and resource managers. The ecologist/resource manager communication channel is central to all scientific aspects of succession management.

22 Obtaining information on succession

knowledge of the local flora or the assistance of a competent plant taxonomist. More complicated studies involving the testing of various management practices or the mathematical analyses of plant community data require expertise in experimental design and statistics and thus should be left to ecologists with graduate research degrees.

2.4 ON-SITE DEMONSTRATED RESEARCH

Resource managers initiating research programmes in parks, nature reserves, natural areas, or on public lands should not ignore the potential value of demonstrated research in the education process or in the arena of public opinion. Demonstrated research occurs when plots or activities associated with research are made available for public inspection while interpretive information is provided (Figure 2.2). Egler (1975) maintained that demonstrated research on succession management is done too infrequently and could be successfully used to influence public opinion regarding various management activities.

Relatively simple modifications are required to transform a research project into a guided tour of succession or succession management.

Figure 2.2 Interpretive aid in the Daniel Boone National Forest, Kentucky, used to explain a management practice and the successional processes involved.

With the proper assemblage of permanent plots and development of interpretive aids, visitors can get a sense of succession that is difficult to duplicate in books or diagrams alone.

2.5 METHODS TO STUDY SUCCESSION

There is a variety of methods that can be used to study succession and to test the effects of various management practices on succession. All of these methods involve assumptions; some involve more than others. Methods discussed in this chapter – with the exception of permanent plots – are not well suited to examining management effects on succession.

2.5.1 Permanent plots

Permanent plots are areas of land where measurements on plant species importance can be repeated over long periods of time. These plots provide the best data for the interpretation of succession and the testing of management practices if the experiments are properly designed (Austin, 1981). Barbour *et al.* (1987), however, stated that 'the establishment and initial sampling of such plots takes a large measure of unselfish forethought'. Few people are willing to begin a scientific experiment where the results will arise only many years in the future.

Resource managers who control disturbance and management activities over large areas of land are ideal individuals to establish permanent plots and then cooperate in permanent-plot research. They are familiar with the research area and can select suitable sites for plot placement. They oversee other types of management activities that could accidentally destroy or disturb vegetation within the plot boundaries. Lastly, resource managers often remain in contact with a single area over long periods of time. They can observe and record chance phenomena that may affect trends within the plot boundaries.

Ecologists establishing permanent plots on private or public lands without the cooperation of a resource manager are destined to fail. Such plots will eventually succumb to human-generated disturbance whether through simple trampling, the cut of a bulldozer, or the senseless activities of vandals. Establishment of 'permanent plots' in unprotected urban or semi-urban areas, however, does provide frustrated ecologists with a new appreciation of repeated human-generated disturbance as a major factor controlling the status of present-day plant communities.

1. Permanent site analysis
Random samples are taken from permanent sites.

☐ - initial samples
▦ - subsequent samples

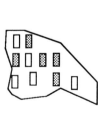

2. Permanent plot with random subsamples
Random subsamples are taken within the boundaries of a large permanent plot.

■ - initial samples
☐ - subsequent samples

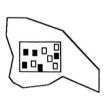

3. Contiguous permanent plots
The entire array of plots are analyzed.

4. Contiguous permanent plots
At replicated sites.

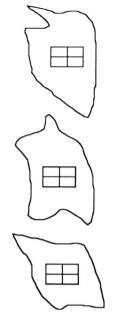

Figure 2.3 Various designs of research involving permanent sites or permanent plots. In all cases repeated measurements are taken through time.

When establishing permanent plots, a primary consideration is how the plot boundaries will be marked. Few studies give information on this, but it is a common reason for plot loss. Boundary markers must be durable enough to last for many years. There is no defensible reason why permanent plots established now should be constructed with a planned lifetime of less than 50 years.

Steel surveyors' stakes with embossed numbers are ideal for marking plot corners but steel reinforcing bars driven into the ground will also work. It is, however, more difficult to establish identifying numbers on the steel reinforcing bar since most tags one would attach to these bars are non-permanent. Wooden fence posts treated with preservatives will not last as long as steel bars but they are easier to locate in vegetation. Unfortunately, people have a tendency to steal wooden fence posts. Buried steel or magnetic objects that can later be located with metal detectors represent the ultimate solution to scientific vandalism. In forests, trees themselves can be tagged and marked to serve as plot boundaries. Large rocks or concrete blocks will also work but they should be large enough so that one person cannot move them. Any type of flagging or plastic stakes should be avoided because these are easily destroyed or dislodged.

Research based on repeated measurements of a single site through time can be designed in a number of ways (Figure 2.3). In many instances, permanent plots are not established but researchers can estimate successional changes by re-analysing vegetation at a permanent site (i.e., permanent site analysis). This is less valuable than research where the actual boundaries of permanent plots are marked. There are few standardized guidelines for how permanent plot experiments should be designed but Austin (1981) provided some suggestions. Repeated measurements taken from entire permanent plots (e.g., contiguous permanent plots in Figure 2.3) are preferred to measurements taken from randomly placed subplots within a larger permanent plot (e.g., permanent plot with random subsamples in Figure 2.3). This experimental design allows better analysis of complex mosaics and invasion fronts. Experiments using permanent plots should be replicated at different sites (e.g., contiguous permanent plots at replicated sites as in Figure 2.3) to recognize the role of chance in vegetation change. Research incorporating experimental manipulation at several levels is preferred to simple description. And lastly, to analyse population-level processes, individual plants or rooted stems should be measured and counted as opposed to estimates of percentage cover (Austin, 1981).

In communities dominated by grasses, herbs, or low shrubs, plots or subsamples can be relatively small. Plots from 0.5 m^2 to 4 m^2 are standard. Permanent plots in tall shrub or forest communities usually are 0.01–0.1 ha.

Specific measurements taken from the permanent plots will vary with the growth forms of individual species. Changing importance of grasses, herbs, and low shrubs is commonly indexed by a measure of percentage cover. This is a fast and efficient technique, but it does not provide data that can be directly interpreted in terms of population changes. Counting individual plants or rooted stems does provide critical demographic information but it is a time-consuming and tedious technique. Entire individuals of large shrubs and trees are usually counted and diameters of the stems measured. Aerial photographs of permanent plots taken repeatedly through time can provide good data on succession although usually a combination of aerial and on-ground techniques are required (Belsky, 1985; Archer *et al.*, 1988).

To demonstrate the power of long-term studies using permanent plots, the results of two studies will be considered, one of old-field succession and the other of secondary forest succession.

Pickett (1982) presented percentage cover data from New Jersey old-field permanent plots. Although there were no demographic data collected in this study, some important points emerge. First, succession was shown to be individualistic in that the percentage cover of individual species appeared to fluctuate independent of other species. Discrete plant communities did not assume importance at discrete points in time. The trends of percentage cover for the different species tended to show broad regions of overlap during 20 years of community development, but the peaks in percentage cover were staggered in time (Figure 2.4). Indeed, many of the species were present in the study plots during nearly the entire study period, thereby supporting the importance of initial floristics composition (Egler, 1954).

While the number of annual species declined and the numbers of perennial herbs and woody species increased during succession, most species did not show classic bell-shaped trends of percentage cover through time. Surprisingly, some species had multimodal patterns of percentage cover. The reasons for this multimodality were not readily apparent.

The trends of percentage cover for a few species could be explained in terms of basic plant biology. For example, the strong first-year dominance of the annual herb *Ambrosia artemisiifolia* followed by decline (Figure 2.4) was explained in terms of seed germination requirements (affecting birth rates) and competitive ability (affecting death rates).

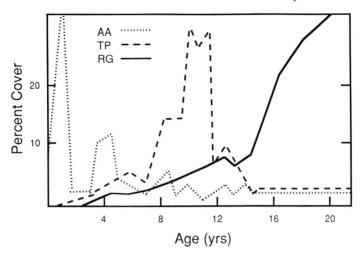

Figure 2.4 Trends in percentage cover of three plant species during 20 years of old-field succession in New Jersey. The species are: AA, *Ambrosia artemisiifolia*, TP, *Trifolium pratense*, RG, *Rhus glabra*. Modified from Pickett (1982). Reprinted by permission of Kluver Academic Publishers.

Pickett's study (actually a study by many people because permanent plots were involved) demonstrated the value of long-term measurements, but many questions remained regarding the causes of succession in this old-field system. Specifically, more information was required on the basic population biology of individual species and on the competitive interactions between species.

Peet and Christensen (1980) presented demographic data on trees collected from permanent plots established in the 1930s and measured periodically until 1978. They were primarily concerned with the roles of mortality (death) and establishment (birth) in the replacement of loblolly pine (*Pinus taeda*) by hardwood species in the North Carolina piedmont.

Plots established in *Pinus taeda* stands of different initial densities and plots established in hardwood forests indicated an exponential decrease in population size of all species over 40 years. Peet and Christensen concluded that such a mortality trend is the dominant factor in population change when major disturbances are absent.

Low level disturbance such as the removal of hardwood saplings in the understorey of pine (Table 2.1) as well as a 1954 hurricane in a hardwood stand (Table 2.2) allowed abundant establishment of hardwood seedlings in the understorey. Severe disturbance such as the

Table 2.1 Effects of different types of artificial thinning on hardwood species invasion of *Pinus taeda* forests in North Carolina

	Tree density (stems ha^{-1})					
	Control		Thinning from below*		Thinning from above†	
Sampling date	1934	1979	1934	1979	1934	1979
Pinus density	2311	355	1047	375	2510	464
Hardwood density	961	2687	190	2737	695	1309

* Small and intermediate size trees removed.
† Largest trees removed.
Data from Peet and Christensen (1980). Used by permission of Kluwer Academic Publishers.

removal of the largest trees in a pine stand (Table 2.1) allowed existing hardwood saplings to emerge into the canopy with little establishment of new seedlings.

Peet and Christensen concluded that pine establishment following severe disturbance will create even-aged stands with little subsequent establishment. In the absence of further disturbance, natural mortality will thin the stands and create gaps that can be filled by suppressed

Table 2.2 Effects of a hurricane on hardwood species establishment in North Carolina hardwood forests

	Tree density (stems ha^{-1})					
	Control			Hurricane thinned in 1954		
Size classes*	1	2	3	1	2	3
Total density 1934	23	76	41	8	55	30
Total density 1950	21	54	46	10	42	33
Total density 1977	65	36	40	335	133	17

* Stem diameter size classes: (1) 2.5–5.0 cm, (2) 5.0–10 cm, (3) >10 cm.
Data from Peet and Christensen (1980). Used by permission of Kluwer Academic Publishers.

Methods to study succession 29

hardwood seedlings and saplings. Severe disturbance during pine dominance can create even-aged stands of hardwoods that also begin to thin as a result of mortality.

These two studies using permanent plots demonstrate clearly that plant communities are dynamic entities. Such long-term observations of single points in space allow researchers to understand both the internal and external forces driving succession.

2.5.2 Chronosequences

If several sites can be accurately aged with respect to time since the last major disturbance, then the plant communities in the different stages of development can be used to piece together a successional pathway. The successful use of chronosequences relies on the assumption that all sites making up the sequence have identical histories of disturbance, biotic influence, and environmental conditions. This may not be valid, especially when the sites are widely separated from each other. The chronosequence technique, however, may be used with a high degree of confidence if many contiguous, different-aged sites can be linked to the same type of disturbance.

The study of succession following glacial advance and river erosion (Fonda, 1974) in the Hoh Valley of Washington State demonstrates the utility of the chronosequence approach. Fonda described a variety of plant communities occurring on river terraces (Figure 2.5), aged the trees, and then pieced together a probable successional pathway. River bars are colonized by red alder (*Alnus rubra*), a stage lasting about 100 years. Sitka spruce (*Picea sitchensis*) and black cottonwood (*Populus trichocarpa*) replace red alder and dominate terraces for about 400 years and then the oldest stages (750 years) of succession are dominated by hemlock (*Tsuga heterophylla*).

2.5.3 Population structure

The historical development and future direction of a plant community can be inferred from an analysis of population age structures. Although this technique is limited to communities dominated by large shrubs or trees that can be accurately aged, it does provide a direct measure of population change through time, an important component of any attempt to describe succession.

By analysing the age structure of living and dead trees in a New Hampshire white pine forest, Henry and Swan (1974) concluded that monospecific white pine stands developed following major disturbance in the late 1700s and have since resisted the invasion of

30 Obtaining information on succession

Figure 2.5 Plant communities on terraces and river bars along the Hoh River in Washington State. A chronosequence technique was used to determine that red alder (*Alnus rubra*) growing on river bars is replaced by sitka spruce (*Picea sitchensis*), visible at the centre of the photograph. Eventually, western hemlock (*Tsuga heterophylla*), visible in the background, dominates after 500 years. Data from Fonda (1974).

hardwood species. Dix (1957), however, noted that a Washington, DC, oak forest was devoid of oak saplings. From the abundance of beech and maple saplings in the understorey, Dix concluded that the oak forest would eventually be replaced by beech and maple.

When observing adult and sapling populations of different species and then inferring successional trends from these observations, it is assumed that the saplings will replace the dominant trees. Such an assumption may not be valid if the saplings do not persist long enough to fill gaps left by dying trees.

2.5.4 Anecdotal information

Often plant community information that is interesting or unique becomes recorded in the journals or memories of people. While such information can be of value in allowing researchers to describe the status of plant communities in the past, anecdotal information seldom

Methods to study succession 31

specifically describes vegetation change through time. As such, successional pathways must be inferred.

Unfortunately, anecdotal information is not data-based and there is no way of determining error rates or accuracy of the information. Regardless of the limitations of anecdotal information, resource managers should realize that complete written records on vegetation change initiated now will be of immense value in the future. These records may be the only information available if and when a succession management plan is implemented.

Written records should include the location of various management activities, the dates of these activities and the condition of the vegetation before management. Repeated observations at yearly or more frequent intervals should record vegetation response to management and measurements of individual species if possible.

2.5.5 Macrofossils and pollen

Long-term changes in plant community makeup are indicated in some special types of ecosystems by the preserved remains of the plants

Figure 2.6 A core of peat drilled from a bog near Fairbanks, Alaska. Preserved plant fragments were used to trace successional development at one site during the last 1200 years.

themselves. In peat bogs or lakes, plant fragments and pollen are deposited and then buried in the accumulating peat or sediments. Identification of these fossils in a stratigraphic section or core can provide direct evidence of vegetation change through time (Figure 2.6). Carbon dating of the material can also establish a relative time frame for these changes.

For example, Jackson *et al.* (1988) removed a core of sediment 140 cm long from a shallow lake in Indiana Dunes National Park. The fossilized remains of plants were identified at increments to determine the successional pathway. These workers found little evidence of succession but instead noted that the time of human settlement corresponded to a decline in many plant taxa and an increase in cattails (*Typha*).

A number of assumptions and difficulties accompany the use of fossil data as a basis for conclusions about succession. Among these is the fact that material from some plant species is better represented in the fossil record than others. Since fossils are an indirect measure of succession, it is difficult to determine causes of vegetation change and of the patterning of succession across the landscape. Plant community changes indicated by palaeoecological data often represent a lengthy time frame that is of little value to resource managers.

SUMMARY

When a natural resource problem involving succession is first identified, a variety of information must be obtained to formulate a succession management plan. Of primary importance is a complete database with on-site descriptions of successional pathways and tests of management activities. Unfortunately, few resource managers have the expertise or time to collect such data. This information, however, can be obtained from published literature or through cooperative research agreements formed with ecologists.

More efficient communication channels need to be established between resource managers and ecologists. On the one hand, resource managers need to relay management goals and problems to ecologists. On the other hand, ecologists need to offer their expertise in experimental design, in data collection, and in the development of alternative management plans. In order to realize their goals, both ecologists and resource managers rely on the expertise and decisions of others, but communication between ecologists and resource managers is central to successful resource management planning.

A number of methods can be used on-site to obtain basic and applied data on succession. Among these are included the analysis of anecdotes, chronosequences, population age structure, macrofossils and pollen, and permanent plots. All of these methods, with the exception of permanent plots, carry numerous assumptions that can weaken the conclusions.

Permanent plots are preferred for obtaining baseline data on succession as well as information on the effects of various management techniques. Measurements in permanent plots, however, must be repeated over long periods of time; few people have historically been willing to initiate and carry out this type of research. Because resource managers oversee activities that could disturb vegetation in permanent plots and also because they maintain long-term associations with the management area, they are the ideal individuals to initiate plot establishment and to cooperate in permanent plot research.

3

Plant populations: growth, decline and persistence during succession

3.1 INTRODUCTION

In a single area of land, as the populations of some species establish and expand, and the populations of other species contract through time, succession results. This view of succession where individual species rise and fall in prominence as opposed to the rise and fall of species assemblages necessitates a clear understanding of the factors causing plant population change. Once succession is interpreted in terms of individual species and their populations, succession management must focus at the population level with management activities limiting or enhancing population growth and decline. This chapter will concentrate on some characteristics of plant populations. It is not by any means a complete treatment of the subject; readers are referred to Harper (1977) for a thorough treatment of it. However, the topics presented here will provide background for understanding the results of management practices considered in later chapters.

3.2 POPULATION TRENDS DURING SUCCESSION

The vast majority of descriptive succession research in herb- or grass-dominated communities does not provide direct measurements of plant demography but rather indices of population changes such as percentage cover or synthetic importance values (Watt, 1960a; Van der Maarel, 1978; Beeftink, 1979; Pickett, 1982). Permanent plots used to monitor population changes in trees are an exception because measurements are usually taken at the level of individual stems or organisms (Peet and Christensen, 1980; Sakai and Sulak, 1985). But even with trees in permanent plots, data limitations can occur if seedlings and saplings are not monitored.

The lack of demographic data on plant populations participating in successional pathways can be traced to the time-consuming nature of such measurements. Another reason can be found in the historical development of succession research. Succession has been closely allied with descriptive plant community ecology, and therefore researchers were concerned with documenting community trends rather than population trends. Once ecologists realize that succession must be approached in terms of individual species and their populations, more demographic data cast in a successional context will emerge. With such demographic data in hand, researchers can then begin to address the complex interspecific and intraspecific interactions (i.e. competitive interactions) contributing to population change.

3.3 POPULATION CHARACTERISTICS AND MEASUREMENTS

3.3.1 Birth and death

Changes in the population size of plant species through time are the result of two opposing population characteristics: the birth rate and the death rate. Immigration and emigration rates, although important determinants of population size in animals, are not easily interpretable in plants. Plants can move into new areas by clonal growth or through dispersal of propagules. But seldom is this immigration associated with a corresponding emigration. In other words, whole plants do not generally move except in unique situations such as the floating of trees in rivers followed by re-establishment downstream.

The time of birth and death in higher animal species is relatively easy to determine, but with plant species the large number of different life cycles (Table 3.1), the frequent occurrence of both sexual and asexual reproduction, and the fact that long-lived plants may contain a large amount of dead tissue creates practical and conceptual research problems. Harper and White (1974) pioneered the broad definition of 'population' – and thus also broad definitions of birth and death – when they considered different levels of plant population structure. The number of individual plants or colonies is one level of population structure whereas the number of repeating units of plant construction (i.e., stems, buds, or leaves) is another level. Depending on life cycles and growth forms of species in a successional pathway, one or several levels of population structure may be targeted in succession management.

Measurement of plant birth and death is limited by the fact that researchers observe and record discrete events in time; these events in nature are continuous. Birth can be considered the point in time when a seedling or a stem emerges from the soil. In a population of dormant buds, birth of a stem may be when a bud is released from dormancy. Death is often indicated by the sudden disappearance of a plant or plant part usually due to herbivory. Wilting, discolouration, or failure to grow are also common indicators of plant death.

Table 3.1 Common life cycles of plant species and variations of these life cycles

Life cycle type	Characteristics and variations
Annuals	Plants that live less than 12 months. Monocarpic species flower and set seed once during a short period of time. Indeterminate species set seed continuously. Winter annuals germinate in the autumn and set seed the following spring.
Biennials	Monocarpic plants that typically grow vegetatively during the first growing season and then flower during the next growing season. Thus they live less than two years. Delayed biennials may live several years before flowering.
Perennials	Plants that live many years. Monocarpic species produce seed once and then die. Polycarpic species produce seeds many times during their lives.

3.3.2 Genets and ramets

When an individual plant has a unique genetic makeup due usually to the fact that it arose from a seed, this plant is considered a genet. If a plant community is composed entirely of annual plants, then each countable individual is a genet. In species capable of asexual reproduction (e.g., root suckers, sprouting from rhizomes, or basal sprouting) the identification of genets becomes difficult and sometimes impossible. A physical connection between plant parts will identify those parts belonging to the same genet. However, when physical connections between plant parts are severed by natural or person-generated forces and the parts are capable of taking up an independent existence, they are then considered ramets. Ramets are typically individual stems or shoots capable of developing a root system. Thus, a population of ramets from the same species in one area may include one or many genets. Population trends of ramets and genets during succession will often be quite different. For example, Hartnett and Bazzaz (1985) showed that the number of *Solidago canadensis* genets invading old-fields stabilized soon after invasion began, but ramet populations continued to increase. When managing a plant species with extensive asexual reproduction, management activities are typically aimed at modifying ramet populations

(Lowday, 1987). Ramet populations are managed by attempting to control rates of sprouting, vigour of sprouting, and also by manipulating the amounts of stored reserves in below-ground tissues.

3.3.3 Age structure

Measurement of ages of individuals in a population reveals population age structure. Age-classes are commonly delimited at yearly intervals and thus are most valuable in the description of population trends in long-lived species. However, age-classes delimited in terms of days could also be used to describe the age structure of short-lived plants or plant parts if an accurate ageing technique is available. The measurement of age structure can reveal past and future population trends.

Two methods are used to determine the age structure of plant populations: observation from the time of birth and observation of annual growth rings or scars. If individual plants are marked and the time of birth is recorded, age structure can then be readily determined at any point in time. More frequently, age structure is determined at one point in time from measurements of growth rings, leaf scars or bud scars; the use of this technique is thus limited to perennial plant species producing such annual growth markers. Ages of individual plants must often be determined directly and destructively because age may not be highly correlated with any other easily measured plant characteristic. In some cases, however, plant size is highly correlated with age and age structure can be determined non-destructively.

Whipple and Dix (1979) identified five types of age-class distributions from a Colorado subalpine forest that, once interpreted, provided information on the history and successional status of tree populations (Figure 3.1). The inverse-J distribution represents a population that is self-perpetuating with a balance of birth and death. Late successional trees will often show this type of distribution. The biomodal distribution represents a population that reproduces in pulses. Such a population may also be self-perpetuating depending on the frequency of the pulses. The decreasing distribution represents a population where birth is not keeping pace with death. Further decreases in birth will lead to a unimodal distribution where a population is not replacing itself. When replacement stops, populations persist to the maximum longevity of the species or until individuals are destroyed by disturbance. Early successional species with no recruitment will show a unimodal age distribution. A random age-class distribution indicates a population that establishes

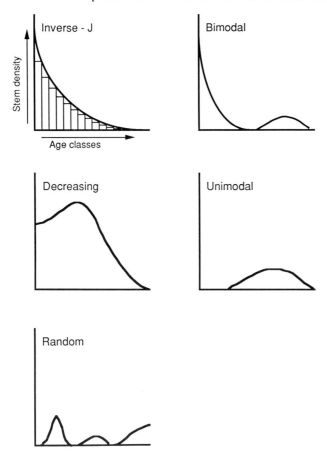

Figure 3.1 Various age-class distributions of stems that can be used to assess past and future population trends. In all figures the *x* axis shows age classes and the *y* axis shows stem densities within the age classes. Modified from Whipple and Dix (1979). Used by permission.

sporadically due to climate change or minor disturbance events. It may also indicate a population just invading a site or one existing in a marginal habitat.

Examination of age structure allows resource managers to determine population trends in target species. If, for example, a species is not showing reproduction and the management goals call for regenerating populations of this species, then disturbances can be designed to meet these management goals.

3.4 POPULATION GROWTH AND PERSISTENCE FROM SEEDS

Harper (1977) identified four roles of seeds in the population biology of plants. First, seeds are the primary means whereby plants colonize new sites; dispersal of seeds to new sites can be aided by a variety of seed designs interacting with wind, water, gravity, or animals (van der Pijl, 1972). Second, seeds, represent a stage in the plant life cycle resistant to unfavourable environmental conditions; seeds can accumulate in the soil as a seed bank. Third, seeds often carry a food supply to nourish the embryo and seedling; the amount of food stored in a seed is an important determinant of seedling success in various types of environments (Salisbury, 1942). Last, because seeds arise from the production of gametes (sperm and egg cells) they contain unique genetic information from the parent plants; genetic variation in the population is maintained.

Much research has been devoted to seeds and their role in population establishment and persistence during succession. Seed availability at a site determines what species can potentially participate in a successional pathway. Furthermore, the history of a site, the disturbance regimen, and the activity of seed-dispersal agents determines seed availability. However, presence of seeds in the soil does not necessarily mean that adult individuals from these seeds will ever exist. Some seeds remain dormant or are destroyed in the soil; others germinate only to be eliminated at some later stage in the life cycle. Changing environmental conditions as a result of disturbance or succession influence the success or failure of plant species to germinate, grow, and produce seed (Holt, 1972; Gross and Werner, 1982). Therefore, a clear understanding of seed ecology is central to many aspects of succession management.

3.4.1 Seed banks

A soil seed bank is due to seed rain from local vegetation, long-distance dispersal of seeds, introduction of seeds by people, and the fact that many seeds express some sort of dormancy in the soil (Figure 3.2). Once seeds enter the soil they can be lost by becoming non-dormant, by predation, by physical destruction, or by decay. Harper (1977) summarized many descriptive data on seed bank sizes and the distribution of seeds in soil. Seed bank sizes range from 122 to 86 000 seeds m^{-2}; seed numbers decline rapidly with depth. Few seeds are found deeper than 15 cm and those that are found beyond this depth are not likely to be viable (Harper, 1977).

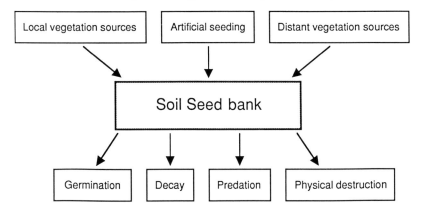

Figure 3.2 Pathways for seed gain to and seed loss from the soil seed bank.

Seed banks can also accumulate above ground. In fire-prone ecosystems such as the shrublands of Australia and some North American pine forests, seeds accumulate in the plant canopy within serotinous cones. The cones open after fire and the seeds are released to the soil. Such an adaptation assures that the majority of seeds will be released when conditions for seedling establishment are optimum (Cowling et al., 1987).

Changes in seed banks during succession are known from a variety of areas around the world. Species in seed banks are often not present in the vegetation; conversely, species in the vegetation may not be present in the seed bank (Oosting and Humphreys, 1940; Livingston and Allessio, 1968; Cheke et al., 1979; Donelan and Thompson, 1980; Jerling, 1983). This experimental result demonstrates that the seed bank is a preserved record of past plant-community status and that fundamental differences exist in abilities of species to regenerate from seed stored in the soil (Grubb, 1977).

Few generalities can be drawn with regard to changes in total seed bank size during succession. Livingston and Allessio (1968) found no trend in seed bank size for 16 sites representing a 70-year successional sequence in Massachusetts. Young et al. (1987), working in a Costa Rican forest, and Jerling (1983), working in a Baltic meadow sere, found maximum seed bank sizes at intermediate stages of succession. Donelan and Thompson (1980), working with a grassland-to-oak-forest sere in Great Britain, and Oosting and Humphreys (1940), working in a North Carolina old-field sere, found decreasing seed bank size with increasing site age.

42 Growth, decline and persistence during succession

Donelan and Thompson (1980) hypothesized that total seed bank size should decline with successional age because species adapted to repeated disturbances produce large numbers of small seeds; in contrast, plants from later successional stages produce fewer but larger seeds. Often among late successional species, no seed production takes place until a critical size or age is reached. Thus seed banks early in succession should be large (Figure 3.3) and seed bank

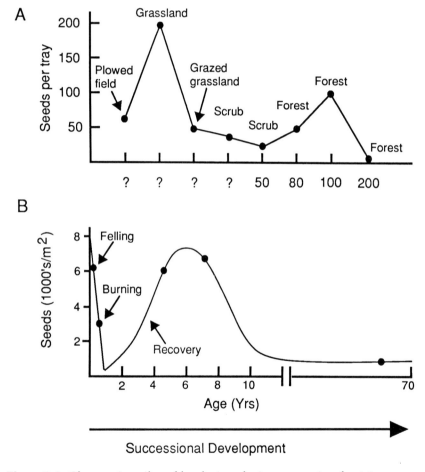

Figure 3.3 Changes in soil seed bank sizes during succession for (a) a sere on English limestone assembled as a chronosequence and (b) a Costa Rican premontane wet-forest sere with points measured directly and patched from a chronosequence. Modified from Donelan and Thompson (1980) and Young et al. (1987). Used by permission of Elsevier Applied Science Publishers and Kluwer Academic Publishers.

size should decline with succession as species replacements occur and the seeds of early successional species are lost from the seed bank (Donelan and Thompson, 1980).

This generalization about changes in seed bank size during succession may adequately describe the sere studied by Donelan and Thompson (1980) (Figure 3.3), but it is not a universal rule. Following tree felling and burning in tropical forests, the soil seed bank is rapidly depleted by seed germination and seed destruction (Figure 3.3). During the next several years the seed bank increases due to dispersal and on-site seed production until a peak is reached at year seven. During the next 60 years the seed bank declines and eventually reaches a size characteristic of early successional stages (Young *et al.*, 1987).

Understanding how the seed bank contributes to plant populations is critical to many aspects of succession management. Later chapters will examine how seed banks in severely disturbed soils can be augmented by topsoiling. In addition, the use of seed banks as a repository for artificially sown seed will be examined as a method of changing both the rate and direction of succession.

3.4.2 Seed germination and seedling establishment

Seed germination requires the presence of a 'safe site', an environment providing the stimuli to break dormancy and also the resources required to allow germination (Harper, 1977). In addition, safe sites are free of the factors that inhibit or destroy the germinating seed. Dormancy itself is the condition where seeds fail to germinate even when placed in a warm, moist environment. Dormancy can be innate, a condition that exists when the seed leaves the parent plant. It can be induced, a condition acquired after leaving the parent plant. Dormancy can also be enforced, a condition created by the lack of some basic resource such as moisture (Harper, 1977).

There is an abundant literature on the innate dormancy of seeds and conditions required to break this dormancy. Seeds from many plant species in temperate regions must experience a period of cold, moist conditions before they germinate, thus insuring that germination does not take place in the same growing season as when the seed was produced. Hard-coated seeds such as those of legumes and a number of non-leguminous mid-successional trees and shrubs are extremely resistant to destruction by animals and physical forces. The seed coat mechanically prevents germination. These seeds persist for long periods of time in the seed bank; dormancy can be overcome simply by nicking or removing the seed coat.

Early successional species, in particular annual plants, produce large numbers of small seeds and form persistent seed banks. Their seeds are stimulated to germinate by high and fluctuating temperatures and light that is not filtered by a plant canopy (Fenner, 1987). These environmental cues that break dormancy also signal the presence of a disturbance, open spot, or gap. Shrub species characteristic of mid-successional stages in forested areas and of mediterranean and heath communities are less easily summarized regarding factors that break seed dormancy. Fire and heat are important factors for many shrub species in chaparral and heath communities (Christensen, 1985); shrub species from forested areas have a hard seed coat that delays germination and may prevent seed destruction during passage through the digestive tract of animals (Fenner, 1987). Canham and Marks (1985) concluded that many shade-tolerant, late successional tree species in the United States and a number of tropical trees have seeds that express little delay between dispersal and germination. These tree species lacking well-developed seed dormancy do not develop a persistent seed bank. Instead, shade-tolerant trees develop a persistent bank of suppressed seedlings that assume dominance when canopy gaps are formed (Fenner, 1987). The complexity of seed response to environmental stimuli suggests that propagule availability and environmental manipulation must be coupled during any controlled colonization efforts.

The availability of safe sites is a necessary component of plant colonization during succession, and so also are the factors controlling seedling establishment and growth to maturity. To determine what micro-sites are favourable for seedling establishment, seeds are often artificially sown into different areas and then their success is monitored. A series of experiments conducted in Michigan old fields (Holt, 1972; Werner, 1975; Gross and Werner, 1982) concluded that the presence of bare ground (litter and vegetation removed) was the microsite most favourable for establishment of various monocarpic herbaceous plant species (Figure 3.4). Litter inhibited germination, and the presence of competing plant species delayed reproduction once plants were established.

Leaf litter affects seed germination by modifying water availability. It forms a barrier between the seed and the soil, thus inhibiting water uptake. This is true for litter laying on the seedbed and also for litter mixed into the seedbed (Naylor, 1985).

Often both litter and standing vegetation can inhibit seedling establishment. For example, Miles (1974) found that seedling establishment in *Calluna* heathland increased with removal of vegetation and topsoil on one soil type, but few effects were evident on

Population growth and persistence from seeds

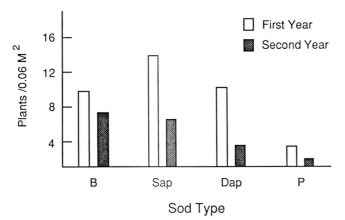

Figure 3.4 Mean number of *Daucus carota* plants resulting from seeds sown in four sod types in Michigan old fields. The sod types were: B, bare ground; SAP, sparse *Agropyron repens*; DAP, dense *A. repens*; and P, *Poa compressa*. Modified from Holt (1972). Used by permission of the Ecological Society of America.

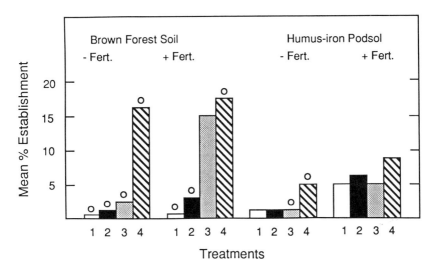

Figure 3.5 Mean percentage establishment of eight different grass and herb species from seedlings in English *Calluna* heath. Treatments were: 1, control; 2, canopy removed; 3, canopy and groundlayer removed; 4, canopy, groundlayer, and upper soil horizons removed. Within each soil and fertilization grouping, bars with circles are significantly ($P < 0.05$) different from successive bars; bars from treatment 4 with circles are significantly different from controls. Modified from Miles (1974). Used by permission.

a different soil type without fertilization (Figure 3.5). Most species had greater establishment when fertilized, but the presence of fertilizer tended to change the effects of the clearing treatments in a complex manner. In heathlands at least, seed weight appears to be an important factor. Small-seeded species are more susceptible to interference from standing vegetation and litter than are large-seeded species.

Tree-seedling establishment during late succession is less well studied than the establishment of early successional species. Again, the availability of bare mineral soil is an important limiting factor. For example, Harrison and Werner (1984) noted that oaks colonized old fields in Michigan only when primary microsites such as bare soil or lichen crusts existed. Walker et al. (1986), working in an Alaskan floodplain sere with five successional states – bare sand, willow (*Salix alaxensis*), alder (*Alnus tenuifolia*), poplar (*Populus balsamifera*), and white spruce (*Picea glauca*) – found that removal of the upper soil layers allowed all tree species in the sere to establish in all successional stages (Table 3.2). In the same study, spruce seedlings showed relatively high survivorship on the plots where the organic matter was removed in comparison to all other species. Spruce is apparently more

Table 3.2 Mean percentages of seeds germinating and establishing seedlings at the end of one growing season for seed of *Salix alaxensis*, *Alnus tenuifolia*, *Populus balsamifera* and *Picea glauca* sown in different successional stages of an Alaskan floodplain

| Species | Treatment | Successional stage | | | | |
		Vegetated silt	Willow	Alder	Poplar	Spruce
Salix	Control	35.5	14.3	—	—	0.2
	Removal*	56.6	17.0	15.5	18.4	16.3
Alnus	Control	14.9	5.7	—	—	—
	Removal	9.7	17.5	12.5	11.8	8.1
Populus	Control	23.8	15.9	0.2	0.8	2.3
	Removal	35.5	23.2	15.5	9.9	28.7
Picea	Control	13.1	12.1	13.3	1.1	0.3
	Removal	26.1	11.8	2.8	8.8	1.5

* Salt crusts were removed in the vegetated silt and willow stages and 01 and 02 soil horizons were removed in all other stages.
Data from Walker et al. (1986). Used by permission of the Ecological Society of America.

tolerant of the low resource availability in the late successional stages (Walker et al., 1986), thus supporting the tolerance interpretation of succession. Designed disturbances that create vegetation openings and bare soil will clearly facilitate seedling establishment of some species. Later chapters will examine how this knowledge can be used to inhibit or encourage target species during succession management.

3.4.3 Seed production and vegetative growth

Because succession involves gradual changes in disturbance regimes and environmental conditions, much effort has been devoted to finding trends in seed production that can be expected during succession. The search for trends has indeed produced some generalizations. For example, a number of studies concluded that early-successional species allocate a greater share of annual production to reproductive structures such as flowers, fruits, and seeds than do species from late-successional stages (Abrahamson and Gadgil, 1973; Hickman, 1975; Newell and Tramer, 1978). In addition, late-successional species have higher root/total biomass ratios than early-successional species (Newell and Tramer, 1978), presumably indicating more vegetative reproduction or below ground storage of energy. Such generalizations of changing sexual reproductive effort during succession are typically based on few species (Abrahamson, 1979) and often it is questionable whether the sites (habitats?) chosen for study actually represent a successional pathway. Instead of searching for a generalized trend in seed production during succession, it may be better to determine how seed production and vegetative growth interact. This has important consequences for management activities especially if one is attempting to eliminate reproduction from seed.

More is known about vegetative growth and seed production in annual plant species than in other types of plants because of the economic importance of annuals as crop plants. Harper and Ogden (1970) grew plants of the annual composite *Senecio vulgaris* under various stress conditions that created a 7-fold difference in plant size. Regardless of plant size, except in extremely dwarfed individuals, plants allocated 21% of their energy to seed production and other structures associated with seed production. Such stability in the allocation to seed is common in annual species where seeds are the sole means of producing another generation. In general, annual species allocate more energy to seeds than do perennial species (Harper, 1977). Thus, sites maintained in an early successional phase may, through time, develop seed banks dominated by annual plant species.

Monocarpic biennials and perennials (plants that live for two or more years and then die after flowering) demonstrate clearly the dependence of seed production on vegetative growth. With most monocarpic species a critical plant size must be reached before flowering is initiated (Baskin and Baskin, 1979; Gross, 1981; Kachi and Hirose, 1985). In addition, the probability of dying or remaining in a vegetative condition is also accurately predicted by plant size measurements at one point in time (Gross, 1981). When monocarpic species are in competition with surrounding plants or when environmental conditions are not favourable, the onset of reproduction is delayed and fewer seeds per plant are produced (Klemow and Raynal, 1985). To maintain viable populations of monocarpic species, designed disturbances must be timed properly so that seed will be provided for a subsequent generation.

Polycarpic perennial species (ones that may flower many times during their lives) are a group of plants that may simultaneously display sexual and asexual production. Seed production in these species tends to lower vegetative growth (Reekie and Bazzaz, 1987).

Assessing the costs of seed production in tree species is complicated by the fact that many north temperate and boreal tree species show 'mast fruiting', the periodic release of large crops of seed (Silvertown, 1980; Sork, 1983). Mast fruiting may be an adaptation for satiating or starving seed predators (Janzen, 1976). It may also be the result of physiological constraints that require a minimum of stored carbohydrate before flowering is initiated (Kozlowski, 1971) or a response to climatic fluctuations (Harper, 1977). Nevertheless, there is evidence to indicate that seed production in trees and some shrubs is size- or age-dependent. Older trees tend to allocate more energy to reproductive structures; young trees allocate more to stem and root growth (Milton *et al.*, 1982; Piñero *et al.*, 1982). Harper and White (1974) found a loose relationship between the life span of trees and the age of first reproduction. For hardwoods at least, the ratio of life span to age of first reproduction was 10 : 1. The age of first reproduction or the magnitude of the current pulse in reproduction are important in succession management because if plants are eliminated before they produce seed or if they are in a low seed production phase, then recolonization may be stopped or limited to seeds arriving from other sites.

Considering tree species that do or do not show mast fruiting, there is abundant evidence that higher seed production is associated with lower vegetative growth and decreased probabilities of survival (Kozlowski, 1971; Gross, 1972; Lloyd and Webb, 1977; Piñero *et al.*, 1982; Luken, 1987a).

3.5 POPULATION CHANGE FROM VEGETATIVE REPRODUCTION

3.5.1 Types of vegetative reproduction

Vegetative reproduction takes three forms of expression in plant species. For some species, vegetative reproduction is the result of vegetative growth that spreads the genet across the landscape. It is the primary method of increasing population size once a seedling is established. In other plant species, vegetative reproduction is expressed only after a disturbance when plant parts are removed or when a plant is fragmented. Both forms of expression involve the formation or release of primordia or buds on various parts of the plant body.

A number of types of vegetative reproduction, referred to as 'reproductive growth', were identified by Tiffney and Niklas (1985). They described two forms of unlinked vegetative reproduction. One is the production of whole plants in axillary meristems such as occurs in *Lemna* and some other aquatic plants. The other is bulbil formation in which bulbils distinctly different from the parent plant in structure are born on the parent plants as in *Lilium*. A third form of unlinked vegetative reproduction is fragment regeneration. Some plants when broken into pieces have the capability of producing adventitious roots and stems. Several willow (*Salix*) species can totally regenerate new plants from pieces of stem tissue (Zasada, 1986) as can a number of aquatic plant species. Whole plants, bulbils, and fragments can be detached and then dispersed much like seeds. Of more widespread occurrence in the plant kingdom and of more concern to managers of succession are the various linked forms of vegetative reproduction (Tiffney and Niklas, 1985).

1. Rhizomes and stolons. Rhizomes (usually fleshy and below ground) and stolons (usually non-fleshy and above ground) are stems that grow horizontally from the parent plant and maintain the ability to develop adventitious roots as well as terminal or axillary shoots. The severing or decay of a rhizome can give rise to independent plants, which in turn can produce their own rhizomes. This type of clonal growth is not typically associated with disturbance, but rather is a successful method of population expansion into new areas.
2. Tillering. This form of vegetative reproduction is found primarily in grasses or sedges. When axillary buds are released from dormancy and vertical growth occurs, a tiller is produced. Each

tiller can develop leaves and roots while remaining attached to the parent clump. Repeated tillering over long periods of time gives rise to a tussock growth form. During disturbance, tillers may be separated from the parent clump and then take up an independent existence.
3. Bud burls and lignotubers. These are structures found primarily at the base of shrub species. They are areas of tissue proliferation where high densities of stem buds are produced from epicormic tissue. Bud burls and lignotubers are common among shrub or tree species prone to frequent above ground destruction by fire (Gill, 1981) and often have a fire-resistant covering. Thus buds on burls or lignotubers are released primarily in response to disturbance, the buds normally remaining dormant. This method of vegetative reproduction is not conducive to horizontal colony spread but large, long-lived clumps can be produced. Fragmentation of a clump does not usually produce a collection of stems capable of assuming an independent existence because the stems are an integral part of the shrub base rather than of the root system.
4. Diffuse basal buds or basal suckering. Diffuse basal buds are not produced in localized areas. Rather they are distributed randomly across the basal zone of shrubs or trees. These buds are produced and released during clump development. Diffuse basal buds may, however, be produced and released in greater numbers in response to disturbances such as fire or clipping. Release of basal buds does not maximize horizontal colony spread, but large, long-lived clumps can be produced. Like bud burls and lignotubers, basal suckering is not conducive to survival of individual stems in the event of fragmentation.
5. Root suckering. Root suckers develop when the cortical cells of roots develop shoot meristems. The spread of a colony by root suckering is thus limited by the horizontal spread of roots. Root suckering is a common form of clonal development in dicots. The best example of clonal spread by this method is found in aspen (*Populus tremuloides*). Extensive root suckering in response to disturbance is also common among dicots and gymnosperms. Survival of root suckers separated from the parent plant possibly results from the fact that each sucker has a root system, but growth is severely reduced by complete severing.
6. Layering. When the lower branches of a plant touch the ground, adventitious roots can be formed at the point of contact. Successive layering can result in clumps of rooted stems. If the rooted stems are separated from the parent plant, they can take up an independent existence. Layering is a common form of vegetative

reproduction in black spruce (*Picea mariana*) a species that often grows in areas of accumulating substrates. Little is known about the fate of layered stems once they are severed from the parent clump.

The distribution of stems of plants across the landscape is strongly affected by the pattern of vegetative reproduction and the subsequent fate of individuals produced by vegetative reproduction. In clonal plants that spread horizontally by rhizomes, stolons, or roots, the guerilla or phalanx growth forms can be identified (Lovett-Doust, 1981). The guerilla growth form has a sparse linear distribution of stems due to little branching of the stolon or rhizome and sporadic stem release (Figure 3.6). The phalanx growth form has stems distributed in a compact front produced by abundant rhizome or stem branching and regular stem production (Figure 3.6). These different types of colony formation can be contrasted with species that display basal or stump sprouting and layering where colonies resemble dense clumps of stems.

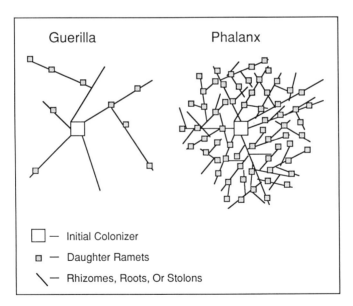

Figure 3.6 Two different patterns of clonal growth in plants. The guerilla pattern is characterized by sparse linear distributions of ramets due to limited rhizome branching. The phalanx pattern is characterized by a compact front of ramets due to intense rhizome branching and regular ramet release. From Lovett-Doust (1981).

3.5.2 Ramet population change

Although population changes of annual or biennial plants without vegetative reproduction have been well studied, the demography of stems, shoots, or tillers arising by vegetative means have been less so. Management activities create situations where community recovery is almost totally dependent on vegetative reproduction. Resource managers are commonly concerned with manipulating the vigour, rate, and pattern of vegetative reproduction. Unfortunately, a major weakness in applied research is the lack of information on how plants with vegetative reproduction respond to management. Nevertheless, some insights can be gained from descriptive research.

In perennial herbs with vegetative reproduction, new shoots may emerge during any month of the year (Noble *et al.*, 1979). But, the majority of new stems emerge from May to October (Noble *et al.*, 1979) and emergence may occur in pulses (Dickermann and Wetzel, 1985). Cook (1985) summarized data on ramet demography of perennial herbs: population dynamics are dominated by ramet birth and death rather than by seedling birth and death; ramet birth and death often coincide, thus leading to population stability; and increases in resource availability increase the flux of ramets by speeding birth and death rates.

Although shrub species and their patterns of vegetative reproduction are quite important to vegetation managers, little is known about the demography of shrubs, especially those that show

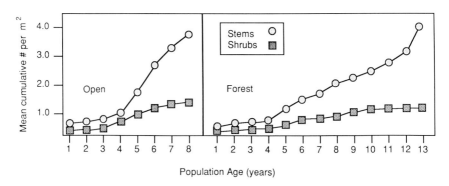

Figure 3.7 Cumulative numbers of stems produced and shrubs established during population development of the invasive shrub *Lonicera maackii* growing in northern Kentucky. Data points are based on five open-grown populations and six forest-grown populations sampled in late July 1986. Modified from Luken (1988) and Luken (unpublished data).

basal stem sprouting. In the naturalized shrub Amur honeysuckle (*Lonicera maackii*) growing in northern Kentucky, the number of shrubs (genets) stabilized 7–10 years after invasion. Demographic trends of stems were habitat specific. Unimodal age-class distributions of stems in open-grown populations indicated maximum stem release from the shrub bases 3–5 years after establishment and then a gradual decline in stem release during subsequent years (Figure 3.7). Few dead stems were found in the open-grown populations. Forest-grown populations had right-skewed or bimodal stem age-class distributions and high numbers of standing dead stems indicating a high flux of stems but also expanding stem populations (Figure 3.7). It was hypothesized that the continued release of new stems in forest-grown populations may be a mechanism that allows rapid shrub response in the event of an opening in the tree canopy (Luken, 1988).

Table 3.3 Comparisons of forest- and open-grown *Lonicera maackii* populations before clipping and 1 year after clipping. Means are presented ± SD

	Forest ($n = 6$)	Open ($n = 5$)
Live shrub density (number m^{-2}) before clipping	0.9 ± 0.2	0.7 ± 0.5
Live shrub density (number m^{-2}) after clipping	0.5 ± 0.2	0.6 ± 0.3
Shrub density percentage change	-42.2	-14.1
Live stem density (number m^{-2}) before clipping	4.1 ± 0.9	3.8 ± 2.9
Live stem density (number m^{-2}) after clipping	5.9 ± 3.4	6.3 ± 3.3
Stem density percentage change	$+45.1$	$+68.1$

From Luken (unpublished data).

Because stem demography in *Lonicera maackii* was habitat specific, it was also hypothesized that shrub response to clipping would be habitat specific. Indeed, a larger number of the shrubs in forest habitats died as a result of clipping, and regrowth in terms of biomass was less in the forest-grown populations (Table 3.3). However, in all populations regardless of habitat, clipping released large numbers of basal stem sprouts so that total stem density, regardless of habitat, increased after clipping. Clearly, bud release in response to biomass removal is a successful adaptation for securing the position of woody plants in disturbed communities.

54 Growth, decline and persistence during succession

3.5.3 Ramet integration

Although the longevity of connections between ramets in clonal plants varies considerably (Pitelka and Ashmun, 1985), there is abundant evidence indicating that when ramets are attached to each other they show various degrees of physiological integration or dependence. When Hartnett and Bazzaz (1983) severed ramets of *Solidago canadensis* from the parent genet, the severed ramets showed lower growth rates and reduced reproduction as compared to attached ramets. Indeed in most clonal plants that have been studied, movement of resources is from old ramets to young ramets. This dependence, best developed early in the life of a ramet, declines with ramet age (Pitelka and Ashmun, 1985).

The net effect of physiological integration among the ramets of a clone is to allow a genet to spread successfully throughout a heterogeneous environment (Hartnett and Bazzaz, 1985). Plants that arise from seed and do not spread vegetatively must exploit the immediate soil environment regardless of its quality. Clonal plants, however, when spreading into new soil environments, are receiving resources translocated from ramets in other soil environments. Thus patchiness in resource availability is homogenized. Such a method of population increase is extremely successful especially in mid- and late-successional environments where establishment from seed is rare or non-existent.

Clonal integration is also an effective adaptation for recovery from disturbance such as clipping or herbivory. In the event that plant parts are removed, resources can be shunted from older stems or from below ground structures, and regrowth quickly occurs. Therefore, during succession management, two components of the propagule pool are of importance: seedbanks and vegetative plant parts. If safe sites for seedling establishment are not present after management activities, vegetation recovery will be dominated by vegetative reproduction. Resource managers attempting to eliminate species with vegetative reproduction must concentrate their efforts on the factors controlling bud release.

3.6 POPULATION DECLINE BY DEATH

3.6.1 Longevity

Although plants can die at any age, and indeed many do die during the seedling and juvenile stages, all plant species have a normal period of longevity. Harper and White (1974) collected data on longevities of a number of perennial species and the results can be summarized: perennial herbs, 2–200 years; shrubs, 30–100 years; coniferous trees, 50–500 years; angiosperm trees, 75–1200 years. Harper and White pointed out that plants with clonal growth may never die; they cited several examples of extremely long-lived clones. The absence of complete senescence in clonal species appears to be an adaptation for perpetual site occupation, a fact that makes clonal plants ideal candidates for development of stable plant communities.

3.6.2 Survivorship patterns

A variety of survivorship patterns are known for various plant species. Not surprisingly these patterns within a species appear to change during population development and in response to environmental conditions. Klemow and Raynal (1983) followed the survivorship of five cohorts of the annual plant *Erucastrum gallicum* over 5 years in a New York State limestone quarry (Figure 3.8). In a year with abundant rainfall a type I (convex) survivorship pattern was observed; in a year that alternated between rainfall and drought a type II (linear) survivorship pattern was observed; and in a year with an early spring drought a type III (concave) survivorship pattern was observed (Figure 3.8). Ramets of clonal plants may show type I or type II survivorship patterns depending on the stage of clonal development and the availability of resources (Noble *et al.*, 1979). In long-lived plants survivorship patterns may suddenly change. Crisp (1978) showed that semi-desert shrubs of Australia had type III survivorship patterns early in life but then changed to type II patterns later in life.

In contrast to gradual mortality through time, Mueller-Dombois *et al.* (1983) described a novel pattern of mortality among various tree species where entire cohorts of trees appear to die simultaneously. Such a pattern of mortality is not adequately characterized as type I survivorship because young and old cohorts of the same species may show die-back within the same general area. Mueller-Dombois *et al.* (1983) proposed that die-back is caused by a combination of ageing and environmental stress and may be a natural event in the population dynamics of many tree species. Cohort senescence may become an important aspect of forest management especially in light of the fact that tree die-backs are now occurring in both North America and Europe. Of particular interest to succession managers are the environmental factors leading to die-back and the patterns of vegetation development following a die-back event.

Mortality, a common component of all plant communities, should be viewed by the resource manager as a harbinger of change. Efforts

56 Growth, decline and persistence during succession

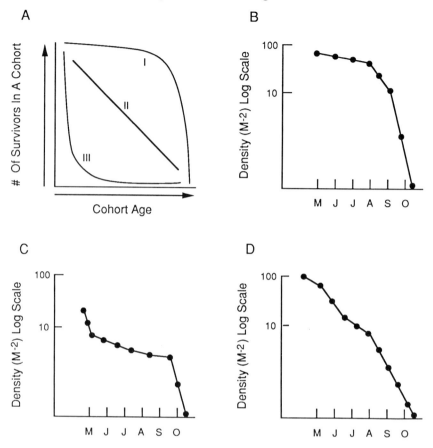

Figure 3.8 Ideal survivorship curves (a) and real survivorship curves (b, c, d) for cohorts of the annual plant *Erucastrum gallicum* growing in a New York limestone quarry. Climatic conditions in different years were: (b) abundant rainfall; (c) spring drought; (d) intermittent rainfall. Modified from Klemow and Raynal (1983). Used by permission.

should be directed at understanding the causes of mortality as well as the species replacements that follow mortality.

SUMMARY

Plant populations expand and contract during succession. Population expansion is the result of seedling establishment, vegetative reproduction, or a combination of these.

Seeds represent an important stage of the plant life cycle producing

Summary

a new generation at a future point in time. Various forms of seed dormancy allow seeds to accumulate in the soil as a seed bank. Seed bank size and makeup both change during succession. Because annual plants produce many small seeds, in some situations seed bank sizes in early successional stages are large and then decline with successional age. Species differences in seed size and germination characteristics may limit the range of sites a species can colonize. In general, the presence of litter, competing vegetation, or organic soil layers have an inhibitory effect on germination and establishment of a wide variety of plant species. Seed production by plants may be constant or periodic, but in many cases a critical age or size must be reached before seeds are produced. Seed production can cause changes in the patterns and amounts of vegetative growth.

Vegetative reproduction is both an adaptation for population expansion and a response to disturbance that allows population persistence. Included among the various forms of vegetative reproduction are fragmentation, root-suckering and layering; and sprouting from rhizomes, stolons, basal buds, and bud burls. Vegetative reproduction produces loose colonies, dense colonies, or clumps. Plant parts produced by vegetative reproduction that can take up an independent existence when severed from the parent are known as ramets. Connected ramets are often physiologically integrated where older ramets provide resources to younger ramets. As the younger ramets age this dependence decreases. Such integration is a successful adaptation for clonal spread into heterogeneous environments. Integration also facilitates rapid growth responses following plant part removal.

Mortality among the individuals of a cohort is usually gradual throughout the lifespan of the cohort. However, stressful environmental conditions will concentrate mortality in the younger age classes. Cohort senescence, or group death, appears to be a novel pattern of mortality in trees that is related to a combination of age and stress.

4

Methods of managing succession: plant and plant part removal

4.1 INTRODUCTION

Succession management is clearly distinct from vegetation management. As outlined in Chapter 1, succession management may involve designed disturbance, controlled species availability, and controlled species performance. Most importantly, succession management recognizes the dynamic character of all plant communities, and this knowledge is then applied in choosing management options. Vegetation management, on the other hand, often ignores developmental aspects of plant communities and instead concentrates efforts on vegetation removal or control to achieve specific but short-lived vegetation 'states'. In many resource management situations, vegetation management could be replaced by succession management with savings in human effort and money over the long term (Egler, 1975).

In some instances the line between succession management and vegetation management is not so distinct. If the goal of a vegetation manager is to arrest succession, then that is a form of succession management. Unfortunately, the vast majority of plant communities on public or private lands are now subject to vegetation management with little concern for alternative management activities, alternative management goals, or for the long- and short-term effects of management on community development patterns. The array of management techniques described in this chapter and in the following two chapters will hopefully stimulate resource managers to explore alternatives that may also achieve traditional vegetation management goals.

Among the techniques used by both succession managers and vegetation managers are cutting, herbicide spraying, fertilizing, burning, grazing, and others. As expected, succession management must often be more precise in the application of management options than vegetation management because target organisms are individual species and their populations rather than simply 'the vegetation'. For example, vegetation managers may opt for blanket herbicide spraying on right-of-way to eliminate all woody vegetation. Succession managers, however, may opt for basal application in an effort to eliminate tall woody species and to release low, woody shrubs.

This chapter will concentrate on management techniques where plants or plant parts are removed by cutting, spraying, burning, or cabling. Both the application of these techniques and the community responses to them will be examined. Techniques altering plant growth rates or dealing with controlled colonization will be treated in later chapters.

4.2 MOWING, CLIPPING AND CUTTING

Mechanical or manual cutting of vegetation is the oldest and presently the most ubiquitous form of succession management. The immediate effect of cutting is reduction in the stature of vegetation. While cutting is often likened to grazing by animals there are important differences between the two. Cutting by mechanical means is commonly non-selective while grazing is selective in regard to species and plant parts. Cutting results in the total removal of aboveground biomass or the cut material is left in place. Grazing removes the cut tissue but organic matter and nutrients are returned in faeces and urine. Lastly, cutting occurs periodically but grazing pressure is periodic or constant.

Regardless of whether cutting is selective or non-selective, the responses of plants to top removal and the responses of surrounding plants when the tops of neighbours are removed are critical to further succession. Differential abilities of species to utilize stored carbohydrate in structures at the soil surface or below ground are important in community development after clipping. In addition, cutting may involve a radical change in the population structure of plants. This is especially true for plants with vegetative reproduction because cutting tends to reduce apical dominance and also releases dormant buds on residual plant structures. Plant size may be reduced and thus also seed output. Lastly, cutting can suddenly change the availability of resources to other plant species, thereby allowing seedling establishment or increased growth.

4.2.1 Grassland and meadows

Grassland and meadows, whether conserved as natural communities or used for the production of biomass and livestock, are communities of low stature where cutting is commonly used as a management technique. Surprisingly, few long-term studies are available where the effects of different cutting regimens are examined in a successional context.

One of the best studies (Wells, 1971) was carried out in English chalk grassland over a period of six years. Plants were cut near ground level, and the cut plant material was removed. Wells' research clearly documented that species respond individualistically to different cutting treatments (Figure 4.1). For example, on deep soils the dominant grass *Zerna erecta* showed a gradual decrease in cover when cut once in spring or once in autumn and a gradual increase in cover when cut only in summer or when left uncut (Figure 4.1). The co-dominant grass, *Festuca ovina*, when growing on medium soils showed gradual decreases in cover under the same cutting regimen as *Zerna*. Herbaceous species also gave mixed results. *Proterium sanquisorba* decreased in cover under all cutting regimens, but *Hippocrepis comosa* increased in cover. *Zerna erecta* was an important keystone species in this community. When allowed to grow uncut it eventually excluded less competitive species.

Both timing and frequency of cutting can have marked effects on community development in grasslands. If tillering of a few grass species is stimulated by cutting, then a decrease in species diversity may result as the vegetative reproduction crowds out plants establishing from seed. This is especially true of cuts that occur in early

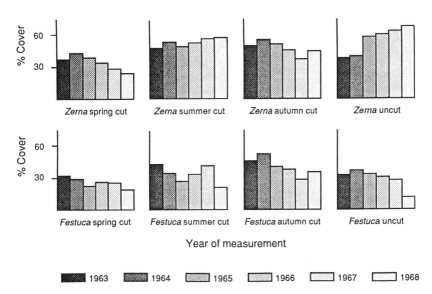

Figure 4.1 Changes in percentage cover of two grass species (*Zerna erecta* and *Festuca ovina*) in English chalk grassland during six years of different clipping regimens. Percentage cover was measured in July. From Wells, T.C.E. (1971).

spring when seedlings are establishing (Oomes and Mooi, 1981). On the other hand, species diversity may also drop dramatically if grasslands are left uncut (Figure 4.2). Commonly, a few tall, aggressive grass species assume dominance (Wells, 1971; Oomes and Mooi, 1981). The best approach for increasing species diversity is multiple cutting that extends through spring, summer, and autumn (Oomes and Mooi, 1981; Powell *et al.*, 1985). This reduces the dominance of a few species and may also increase safe site availability through litter removal or soil disturbance (Figure 4.2). Less competitive herbaceous species can then establish. All management suggestions for grasslands are site-specific because even the same species on slightly different sites will respond differently to cutting schedules (Vestergaard, 1985).

4.2.2 Heathlands

In the absence of intense management, heathlands of western Europe originally dominated by heather (*Calluna vulgaris*) are invaded by a variety of trees and other plant species. Mechanical cutting has been used to control unwanted species and also as a method of regenerating *Calluna*. Responses of woody heathland plants to cutting will depend on their adaptations for vegetative reproduction and on their physiological condition at the time of cutting.

In general, cutting of invasive tree species such as birch (*Betula pendula*) in an effort to reverse succession is not successful. Resprouting from basal buds quickly regenerates the tree populations (Marrs, 1987a, b). Repeated cutting can improve the rate of kill (Marrs, 1984). For total control, however, cutting must be done in combination with spot herbicide application.

Another invasive species of heathlands is bracken (*Pteridium aquilinum*). Here again, single cuts lead to rapid regeneration.

Figure 4.2 See pages 62 and 63. Effects of varied cutting frequency on community composition of an *Arrhenatherion elatioris* grassland in The Netherlands after 14 years of management. (a) With no cutting *Carex riparia* dominates. (b) With cuts in May and September the community remains diverse with 6–10 species. (c) Cutting in May, July and September causes a strong community shift. *Agrostis stolonifera* and *Ranunculus repens* dominate. (d) Cutting in August only causes the community to be gradually dominated by *Alopecurus pratensis*. (e) When cut in July and September, the community remains diverse. (f) When fertilized with 50 kg N, 20 kg P and 20 kg K per ha the community is dominated by *Alopecurus pratensis* and most other species disappear. Photographs courtesy of Dr M.J.M. Oomes.

Mowing, clipping and cutting

Table 4.1 Height, density, and above ground biomass of bracken on 12 November after receiving single cuts during spring and summer of the same year

Date of cutting	Height (cm)	Density (fronds m^{-2})	Biomass (g m^{-2})
Control	149.5	25.5	445
6 June	130.9	28.2	525
20 June	125.9	25.9	354
4 July	92.1	25.2	268
18 July	74.3	17.9	148
1 August	53.4	6.0	74
15 August	46.3	1.5	14

Modified from Lowday et al. (1983). Used by permission of Academic Press.

However, Lowday et al. (1983) found that regrowth was less vigorous when cuts were made after mid-July. In all cutting treatments except the late August cut, frond density increased the following year due to frond bud release (Table 4.1). A single cut in late August removed the maximum amount of biomass and also gave less resprouting the following year (Lowday et al., 1983). If cutting is used to control species performance, then the point of maximum biomass must be determined and cuts should be timed to remove this biomass before translocation to below ground storage organs occurs.

Cutting in heathlands is also used to regenerate desirable heathland species such as *Calluna vulgaris*. This is a successful approach provided the stands are not old and senescent. When stands of *Calluna* are senescent, cutting will not lead to a flush of bud release but safe sites for the establishment of seedlings may be inadvertently created (Marrs, 1987b).

4.2.3 Semi-desert scrublands

Cutting is used as a method to control the performance of trees or shrubs in areas where grasses and herbaceous species are preferred for their grazing value. Large woody plants tend to reduce the standing crop and diversity of grasses and herbaceous plants, but this effect – as in the case of eastern redcedar (*Juniperus virginiana*) growing in Oklahoma grassland – is limited to the area immediately under the tree

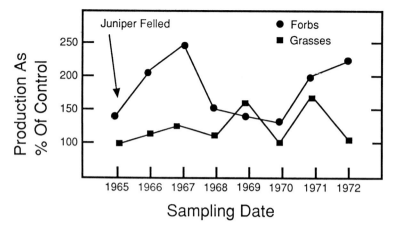

Figure 4.3 Production responses of grasses and forbs to the cutting of alligator juniper (*Juniperus deppeana*) in north-central Arizona pinyon–juniper woodland. Modified from Clary (1974). Used by permission.

canopy (Engle *et al.*, 1987). Some range managers maintain that grass production can be increased by removing these trees.

Available data do not strongly suggest that cutting of junipers from rangelands has a predictable effect on grass or herbaceous species. For example, in Arizona pinyon-juniper rangeland when alligator junipers (*Juniperus deppeana*) were cut (Clary, 1974), perennial forbs showed an increase in production but in three of the seven years there was no measureable response (Figure 4.3). A positive grass response was not measured until four years after cutting (Figure 4.3). Long-term changes in grass and forb production after tree removal appear to be linked to climatic fluctuation rather than to tree removal. Moreover, it is not known how cutting affects the release of juniper seedlings or juniper seedling establishment.

4.2.4 Forests

Management of forests by various cutting regimens falls under the domain of silviculture. Silviculture is an advanced form of succession management with sophisticated techniques and principles that could also be applied to communities other than forests. In addition, silvicultural techniques could be used to achieve management goals other than maximum wood production.

A large number of cutting strategies are used in forests. Regeneration cuttings remove old individuals and provide environmental conditions suitable for establishment of new individuals. Intermediate cuttings done during stand development improve growth rates or growth forms of remaining trees. Intermediate cuttings include thinning, release cuttings, and pruning (Smith, 1986).

Forest harvest methods include: clear cutting where all trees in a single area are removed; the seed-tree method where most trees except a few are left standing to provide seed for stand regeneration (Figure 4.4); the shelterwood method in which partial cuttings are made during the development of a stand; and the selection method where mature trees are removed in an effort to create uneven-aged stands. The development of plant communities under the pressure of periodic harvest is known as secondary succession. Secondary-succession pathways are determined by the harvest method, the availability of tree seeds, the substrate condition, the ability of trees to resprout from cut stumps, and the response of understorey plants to canopy removal.

Tree seed availability will vary with the timing of cutting. Cutting in winter assures that seeds are dispersed and available for stand regeneration the following year (Zasada, 1986). Seed distribution after cutting is determined by the distribution of seed trees and by the activity of seed dispersal agents. Clear cuts can be reseeded by adjacent vegetation. Wind disseminated seeds, however, will travel only one to five times the height of the adjacent forest (Smith, 1986).

Microsites for germination and seedling establishment are required to insure uniform tree regeneration after cutting. Bare mineral soil is the preferred substrate for seedling establishment of many tree species. Forest harvest techniques create sites favourable for seed germination due to the movement of heavy equipment and the skidding of logs. However, in most cases artificial soil scarification, the removal of slash (Figure 4.4), and the control of understorey species are necessary to insure uniform tree seedling establishment after cutting (Zasada, 1986).

In addition to regeneration from seed, tree sprouting from stumps or roots strongly influences secondary succession (Figure 4.5). Growth rates of sprouts after cutting, at least in the early stages of development, will exceed the growth rates of trees establishing from seed. This is due to the fact that sprouts are supported by established root systems. If sprouting is not controlled after cutting then the community that develops may be dominated by sprouting species.

Sprout vigour can be modified by the timing and method of cutting. Sprouting is inhibited by spring or summer cutting and enhanced by winter cutting (Smith, 1986; Zasada, 1986). This is presumably due to the fact that carbohydrate storage is maximum in winter and at a minimum in spring and summer (Kozlowski, 1971). The abilities of

68 Plant and plant part removal

Figure 4.4 Various methods of cutting and site preparation used in the Daniel Boone National Forest, Kentucky. (a) The seed tree harvest method in a short leaf pine (*Pinus echinata*) forest. Most trees are harvested, but a few veteran individuals are left standing to provide seeds. (b) Recently logged hardwood stand. Saplings and shrubs have been killed with herbicides to decrease competition with seedlings. (c) Clearcut hardwood stand where slash was windrowed and burned to prepare the seedbed. Photographs courtesy of Dr P.J. Kalisz.

most trees to sprout decreases with age. Harvest methods that destroy large portions of stumps, stump bark, or roots will also lead to decreased sprouting (Zasada, 1986).

The number of sprouts arising from stumps is dependent on stump size, but this varies with species. Mroz *et al*. (1985) determined species-specific relationships between stump diameter and the number of stump sprouts. The sugar maple relationship was bimodal, the red maple (*Acer rubrum*) relationship was unimodal, and the black cherry (*Prunus serotina*) relationship was skewed (Figure 4.6). Not surprisingly, black cherry forms a significant component of regenerating forests in eastern US. Relationships between stump diameter and sprout number in part reflect the potential of certain species to dominate communities after cutting. Tree species that lose sprouting ability with age will tend to be eliminated if cut when forests are

Figure 4.5 (a) Soon after some tree species are cut, numerous stump sprouts are released at or near the soil surface. (b) Multi-stemmed clumps of maple (*Acer* sp.) that sprouted after forest harvest in the Daniel Boone National Forest, Kentucky. Species that can form stump sprouts will dominate regenerating communities.

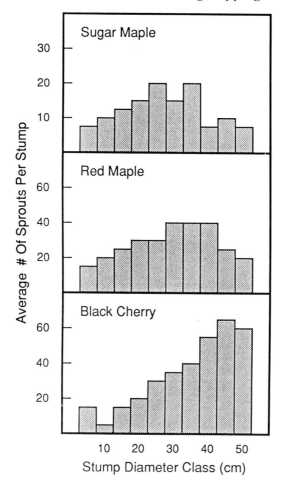

Figure 4.6 Relationships between stump diameter and the number of sprouts per stump for sugar maple (*Acer saccharum*), red maple (*Acer rubrum*), and black cherry (*Prunus serotina*) growing in Michigan. Based on a total sample of 2126 stumps. Modified from Mroz *et al.* (1985). Used by permission.

mature. On the other hand, species that develop sprouting potential with age will assume dominance after cutting of mature forests.

There is perhaps more known about the root sucker response of aspen (*Populus tremuloides*) to cutting than any other tree species. This is because root suckering is the primary method of regenerating aspen stands for wood production. Schier *et al.* (1985) concluded that clearcutting of aspen maximizes root sucker production. Partial

72 Plant and plant part removal

cutting tends to produce fewer suckers that do not perform as well due to competition with uncut stems. Girdling of stems before cutting reduces the number and longevity of root suckers (Schier and Smith, 1979). This is presumably because girdled trees remain physiologically active up to three years after girdling and the stems drain the root systems of carbohydrates. Lastly, disturbance during logging affects root sucker production. Skid trails or roads created during logging remain devoid of aspen suckers as a result of root damage; large accumulations of slash tend to inhibit successful emergence of suckers (Schier *et al.*, 1985).

The resilience of some forest communities to cutting is aptly demonstrated in a study of mixed mesophytic forest communities in eastern Kentucky (Muller, 1982). When an old-growth stand with no recorded history of cutting was compared to a 35-year-old stand that regenerated naturally after cutting, few differences in stand

Figure 4.7 A forest in North Carolina clearcut in 1931. Shrubs and saplings were also cut and burned. Forty years later the stand is dominated by sprouts, primarily of yellow poplar (*Liriodendron tulipifera*), red oak (*Quercus rubra*), and black locust (*Robinia pseudoacacia*) (McGee, 1972). Photograph courtesy of US Forest Service and Dr Donald E. Beck.

composition or basal area were found. Due to a combination of resprouting and seedling release, community recovery was nearly complete. Often sprouts from roots or stumps of cut trees are the long-term dominants of regenerating forests. For example, Figure 4.7 shows a forest stand in the Appalachian Mountains of North Carolina 40 years after clear cutting. Periodically during community re-generation multiple tree sprouts were thinned and now the forest is nearly dominated by sprouting tree species such as yellow-poplar (*Liriodendron tulipifera*) and oaks (*Quercus* spp.) (McGee, 1972). Uhl *et al.* (1982) measured community development patterns of an Amazon Caatinga forest three years after clear cutting. They found that stump sprouts comprised 73% of the above ground biomass after cutting. Thus tree species capable of sprouting were over-represented during secondary succession. Uhl *et al.* (1982) hypothesized that the dominance of stump-sprouting tree species could be used as an indicator of those communities with a history of cutting.

4.3 HERBICIDES

Herbicide use is a management technique for the elimination of undesirable plants or for the elimination of an entire species from one area. While a large portion of present-day herbicide use is done to achieve complete vegetation control with little concern for successional development (i.e., vegetation management), herbicides can play a strong role in succession management where individual species are targeted for elimination and a specific plant community development pathway is desired. Indeed, Marrs (1985a) pointed out that herbicide use in nature reserves is very different from herbicide use in agriculture or forestry. In nature reserves there may be one target species and all others are desirable whereas in agriculture and forestry there may be one crop and all other species are undesirable. Of concern to succession managers who opt to use herbicides is (1) the selectivity of herbicide action, (2) the selectivity of the herbicide application technique, and (3) the responses of target and non-target species to herbicide application.

Norris (1967) identified three types of selective herbicide action in plants (1) different species respond differently to one herbicide; (2) one species responds differently to different herbicides, and (3) plants of one species growing in different areas respond differently to one herbicide. Norris concluded that the selective effect of herbicides, at least for tree species, could be described by the following equation:

$$EFF = INT \times ABS \times TRANS \times STAB \times TOX$$

74 Plant and plant part removal

where:

> EFF = the ultimate effectiveness of one herbicide
> INT = the amount of herbicide intercepted by plant leaves
> ABS = proportion absorbed
> TRANS = proportion of absorbed material which is translocated to the roots
> STAB = proportion of material in the roots in an active form
> TOX = ratio of toxicity to root tissue as compared to other herbicides

Woody plant species are the most common targets of herbicide use in succession management. The effectiveness (or lack thereof) of a herbicide is expressed as leaf kill, leaf and stem kill, or root kill. Egler (1975) maintained that root kill should be the goal in all succession management. But woody plant species (with the exception of most conifers) will resprout if the tops are killed and the roots are not. Achieving root kill of trees and shrubs with herbicides is often difficult, and this task is made more difficult when woody plants occur in a matrix of non-target species. Although herbicides are manufacturer-tested and certified to work on specific species, often the response is variable due to environmental conditions and differences in plant vigour. Screening trials carried out in the field are necessary to determine the best herbicide and the best method of application before herbicides are used in general practice (Marrs, 1985a).

Increasingly strict regulations regarding the toxocity and secondary ecological effects of herbicides have limited the numbers of new herbicides coming on the market since the 1950s and 1960s. However, new methods of herbicide application which are constantly being developed, strive to deliver higher selectivity and lower environmental contamination. More selective use of herbicides in succession management is being demanded by the general public as concerns about environmental pollution and the effects on human health become widespread (Pendleton, 1983).

Table 4.2 lists a variety of herbicide application methods that can be used to control woody plant species. They are ranked (1–16) from the least selective method at 1 to the most selective method at 16. Obviously, methods in which herbicides are sprayed to foliage or scattered on the ground are less selective than methods in which herbicides are applied to stems or stumps. When spraying is used there is always the possibility of drift and effects on non-target species. Highest selectivity is achieved when notches or frills are made in the tree bark and then filled with herbicides.

Table 4.2 A relative ordering of various herbicide application methods (1 least specific and 16 most specific) with regard to specificity of herbicide application on target tree species

1. Broadcast foliar application by aircraft
2. Broadcast soil application by aircraft
3. Broadcast foliar application by truck or tractor
4. Broadcast soil application by truck or tractor
5. Foliar application by hand sprayer
6. Broadcast bark or stem spraying
7. Stump spraying
8. Stump painting
9. Basal bark spraying
10. Basal bark painting
11. Rope or wick application
12. Bark frill and spray
13. Bark frill and paint
14. Bark notch and spray
15. Bark notch and paint
16. Notch and inject

4.3.1 Application methods and responses of target species

As indicated by Norris' (1967) equation to describe herbicide effectiveness, numerous factors are involved in determining the ability of a herbicide to achieve root kill. The weed-science literature is replete with research on the killing success of herbicides applied by various methods. Resource managers are referred to this literature (especially the journals *Weed Science* and *Weed Research*) to determine if a specific herbicide has been tested on the target species they are interested in controlling. Table 4.3 provides the common names, trade names, and manufacturers for most herbicides now used in succession management. An attempt will be made here to summarize some of the general results regarding target species/herbicide interaction.

Small trees and shrubs show vastly different tolerances to the same herbicide. For example, Cantrell *et al.* (1986) achieved complete root kill of turkey oak (*Quercus laevis*) in Florida by injection of 2,4-D, triclopyr, 2,4-D + picloram and haxazinone while dicambra, glyphosphate, and dichlorprop did not give acceptable root kill and resprouting occurred. Timing of herbicide application is critical.

Table 4.3 Common names, trade names and manufacturers of herbicides frequently used in succession management

Common name	Trade name	Manufacturer
Ammonium sulphamate	Ammate X-NI	Dupont
Asulam	Asulox	May and Baker
Bromacil	Hyvar X	Dupont
Dicambra	Banvel CST	Velsicol
Dichlorprop	Weedone	Dupont
Glyphosphate	Roundup	Monsanto
Hexazinone	Velpar L	Dupont
Krenite	Krenite-S	Dupont
Monuron	Telvar	Dupont
Picloram	Tordon	Dow
Tebuthiuron	Spike	Elanco
2,4-D	2,4-D	many
2,4,5-T	2,4,5-T	many
2,4-D + picloram	Tordon RTU	Dow
Triclopyr	Garlon 3A	Dow

Greater kill is often achieved if herbicides are applied at the time of maximum physiological activity. For temperate regions this is spring.

Large trees pose a unique challenge to succession managers because of difficulties in delivering adequate amounts of herbicides to the roots. For example, Yeiser (1986) concluded that stem-injected glyphosphate/2,4-D mix, picloram + 2,4-D, or hexazinone gave better kill of oaks and hickories in Arkansas than did dicambra. Oaks were killed more easily than hickories; large hickories were extremely resistant to herbicide treatment. In large trees, the best time for herbicide application is species-specific. Striped maple (*Acer pensylvanicum*) control was best when sprayed in July–September; beech (*Fagus grandifolia*) control was best when sprayed in August–October (Horsley and Bjorkbom, 1983). The timing of tree cutting can influence the response of stump sprouts to herbicides. Stumps of trees cut in the dormant season were killed most effectively by treatment with picloram + 2,4-D, hexazinone, or dicambra; stumps of trees cut during the growing season were killed by treatment with triclopyr or glyphosphate (Zedaker *et al.*, 1987). From the variability in tree response to different herbicides and application methods it is clear why screening trials are necessary before a large-scale spraying programme is started.

4.3.2 Herbicides in rights-of-way

Succession management on rights-of-way by using selective herbicide application is perhaps the best example of how controlled species performance can produce a relatively stable, treeless plant community. Utility rights-of-way exist to accommodate above ground transmission lines. Trees can interfere with line function and they make line servicing difficult. Rights-of-way must therefore be kept free of trees and must also be accessible to line maintenance crews.

The management of rights-of-way, at least in the US, has changed dramatically through time. When herbicides first became available, many thousands of hectares of rights-of-way were blanket sprayed with herbicides to achieve complete brush and tree control. Indeed, blanket spraying and brush cutting are still common management techniques in some parts of the country. In forested areas, however, these traditional management practices created much bare ground that facilitated tree regeneration by seeds or sprouts. Many rights-of-way communities several years after spraying had more tree stems than before spraying (Byrnes, 1961). Cutting of trees or brush also produced communities that were totally composed of tree sprouts. These management activities had to be repeated often in order to keep trees out of the transmission lines.

A novel management goal for rights-of-way was examined after researchers observed that certain shrub or herbaceous species could successfully resist tree invasions for many years (Pound and Egler, 1953; Niering and Egler, 1955; Niering and Goodwin, 1974; Niering *et al.*, 1986). Selective herbicide use played a large role in achieving this goal (Egler, 1949). Instead of concentrating on short-term solutions to the problem of tree invasion, methods were sought to encourage specific plant communities that would resist tree invasion through competitive interactions.

One such approach for achieving stable, low shrub communities in rights-of-way was tested in a recent study in Connecticut by Dreyer and Niering (1986). They compared succession in rights-of-way after two different methods of herbicide application to trees: stem foliar spraying and basal spraying. The stem-foliar sprayed communities had more herb coverage and less shrub coverage than communities where trees were basal sprayed (Table 4.4). The stem-foliar spraying was semiselective, and herbicide drift eliminated shrubs near the trees.

Table 4.4 Comparisons of percentage shrub, herb and tree cover on Connecticut rights-of-way three to seven years after herbicide treatment of trees by basal or stem-foliar methods

Research site	Shrubs	Herbs	Trees
Basal sprayed R-O-W			
Connecticut Arboretum	95	20	5
Flanders	68	60	4
Ledyard North	80	65	7
Basal R-O-W (mean, $n = 3$)	81	47	5
Stem-foliar R-O-W			
Ledyard South	50	85	7
Middlefield	30	85	7
Stem-foliar R-O-W (mean, $n = 2$)	40	85	7

Modified from Dreyer and Niering (1986). Used by permission of Springer Verlag Publishers.

Herbicide drift effects on non-target species can work against management goals. Specifically, it is the low shrubs such as greenbrier (*Smilax rotundifolia*), oriental bittersweet (*Celastrus orbiculatus*), mountain laurel (*Kalmia latifolia*), and huckleberry (*Gaylussacia baccata*) that best resist tree invasion. In the Connecticut study, tree seedlings were less frequent in areas of high shrub coverage, and rights-of-way treated by basal spraying had fewer tree seedlings (Table 4.4). The ability of shrubs to resist tree seedling establishment may be the result of competition for resources such as light, water, or nutrients, the production of allelopathic chemicals by shrubs, or intense grazing of tree seedlings under the shrub canopy (Dreyer and Niering, 1986). Regardless of the mechanism of tree inhibition, the results of this study clearly demonstrate the value of selective herbicide

---

Figure 4.8 Powerline rights-of-way in Pennsylvania after 31 years of succession management. The goal is to develop a low shrub–herb community under the wires. (a) Management included cutting by hand in 1951, stem-foliar spraying of trees in 1953, selective basal spraying of trees in 1966, and stem foliar spraying of trees in 1982. (b) Management included cutting by hand in 1951, broadcast foliar spraying in 1953, selective basal spraying of trees in 1966, and stem foliar spraying of trees in 1982. A recently sprayed tree is visible in the foreground. Photographs courtesy of Dr W.R. Byrnes.

application as a method of eliminating some species and encouraging others during succession.

Bramble and Byrnes (1976, 1983) documented vegetation development on Pennsylvania rights-of-way during 30 years of various herbicide treatments (Figure 4.8). They found that blanket spraying, basal spraying, and cutting without spraying all produced distinctly different successional pathways. Cutting without spraying caused the smallest change in species composition and successional trajectory. A few years after cutting, the original forest community was regenerated except that trees were replaced by tree sprouts and few new tree species entered the community. Blanket spraying eliminated most of the trees, and a sedge–herb–grass community was established. This community then slowly began to be invaded by shrubs, tree sprouts, and tree seedlings. Shrub importance was re-established about 15 years after spraying. Basal spraying produced a bracken–sedge–herb–shrub community that also began to be invaded by tree seedlings after six years. Most notable was the preservation of the shrub component in the basal-sprayed communities. Bramble and Byrnes concluded that selective follow-up basal spraying of trees was required on nearly all types of plant communities in their study sites to maintain a tree-free condition. Differences in follow-up treatments produced different successional pathways. Less selective herbicide application techniques typically worked against the goal of preserving the low shrub component.

On Pennsylvania rights-of-way, monospecific communities of herbs and shrubs differed dramatically in their ability to resist tree invasion (Table 4.5). For example, low early blueberry (*Vaccinium angustifolium*) or meadow fescue (*Festuca elatior*) were quite successful at resisting tree invasion, but stands of sweetfern (*Comptonia peregrina*) and poverty grass (*Danthonia spicata*) were not (Table 4.5). Tree species that did successfully invade these rights-of-way were black cherry (*Prunus serotina*), red maple (*Acer rubrum*) and sassafras (*Sassafas albidum*). Tree invasion was most obvious where seed sources of these trees were adjacent to the study sites.

Succession research on rights-of-way in the eastern US suggests that herbaceous or shrub species from the local floras can be managed by selective herbicide application to produce relatively stable, treeless vegetation. The best plant species to resist tree invasion – and thus those that should be protected from spray – are low, aggressive species that rapidly establish complete soil coverage. These herbicide-maintained communities apparently satisfy management goals and may in the long run be cheaper to maintain than communities that are blanket sprayed or cut on a frequent rotation. Lessons learned in

Table 4.5 Densities of invading tree seedlings (<1 m tall) and numbers of emerging trees (1–2 m tall) in single-species patches on a Pennsylvania right-of-way after various herbicide treatments. These patches occupied 21% of the right-of-way (ROW)

Dominant species	Tree seedlings (no. ha^{-1})	Emerging trees per ha of ROW
Meadow fescue (*Festuca elatior*)	0	0
Rough goldenrod (*Solidago rugosa*)	1 235	7.2
Bear oak (*Quercus ilicifolia*)	2 470	1.0
Early blueberry (*Vaccinium angustifolium*)	3 087	0
Blackberry (*Rubus allegheniensis*)	4 233	16.1
Hayscented fern (*Dennstaedtia punctilobula*)	5 248	4.2
Poverty grass (*Danthonia spicata*)	6 175	1.7
Narrow goldenrod (*Solidago graminifolia*)	8 892	1.2
Sweetfern (*Comptonia peregrina*)	9 880	2.7
Mountain laurel (*Kalmia latifolia*)	9 880	0.2
Witchhazel (*Hamamelis virginiana*)	11 115	1.7
Loosestrife (*Lysimachia quadrifolia*)	12 350	0.2
Late blueberry (*Vaccinium vacillans*)	17 290	0.2
Huckleberry (*Gaylussacia baccata*)	46 930	5.2

Modified from Bramble and Byrnes (1976). Used by permission.

rights-of-way could be applied to other types of situations where treeless vegetation is required.

4.3.3 Rangelands

Large areas of land in western and southwestern parts of the US are managed for the grazing of livestock. On these rangelands the invasion of shrubs and trees is in general considered undesirable because woody species compete with grasses and herbs for moisture and thereby reduce the value of the land for grazing purposes. However, shrubs and trees are not universally detrimental to the growth of grasses and herbs in rangelands. There are situations where dense stands of grasses develop under the canopies of trees and shrubs. This is presumably the result of greater availability of nutrients (Garcia-Moya and McKell, 1970) or moisture (Tiedemann and Klemmedson, 1977) under the

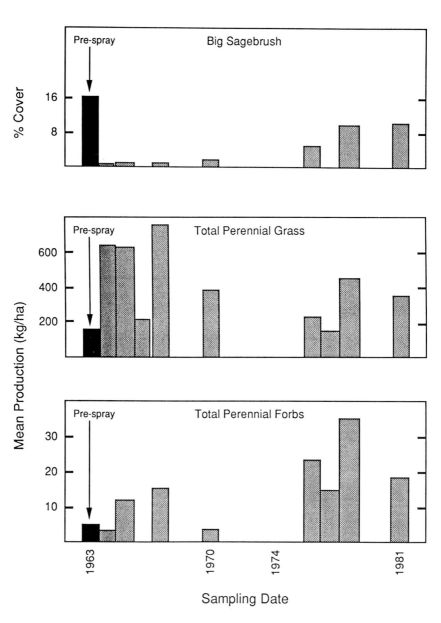

Figure 4.9 Production responses of perennial grasses and forbs following a single herbicide application to big sagebrush (*Artemisia tridentata*) growing in Wyoming. The top panel shows the recovery of big sagebrush populations during the study period. Modified from Wambolt and Payne (1986). Journal of Range Management (1986). Used by permission.

shrub canopies. Not surprisingly, a variety of successional changes have been measured as a result of herbicide treatments to eliminate shrubs and trees from rangelands.

An overwhelming number of studies suggest that grass production is increased by herbicide treatment of the shrub layer whereas forbs are not stimulated. However, even the grass response is short-lived and may last only 4–7 years (Thilenius and Brown, 1974; Jacoby et al., 1983). Few studies in rangelands follow population trends in both target and non-target species after herbicide application. The study by Wambolt and Payne (1986) is an exception. They measured understorey cover, understorey production, and big sagebrush cover over a period of 18 years in Montana rangeland after herbicide (2,4-D) treatment of big sagebrush in 1963. In general, there was greater production of understorey species with shrub removal (Figure 4.9). However, this response was best expressed by perennial grasses, and the response of perennial forbs was less dramatic. Some grass species responded better than others. For example, bluebunch wheatgrass (*Agropyron spicatum*) production was consistently stimulated by spraying whereas Sandberg bluegrass (*Poa sandbergii*) production was not. During this study big sagebrush showed a gradual return of canopy coverage to pre-spray levels (Figure 4.9).

Overall, it appears that elimination of shrubs and trees from rangeland with herbicides does tend to stimulate primarily grass species. However, the response is individualistic and woody plant removal is likely to cause subtle changes in community composition that may not be detected by herbage production estimates that are seldom done to species level.

4.3.4 Heathlands

Using herbicides to selectively eliminate weedy species from heathlands produces both positive and negative effects on further community development. In general, to achieve complete kill of woody plant species a combination of different herbicides is required (Marrs, 1987a). When weedy tree species are killed, the bare patches that remain provide suitable sites for the establishment of *Calluna* seedlings as well as other plant species. For example, *Deschampsia flexuosa* readily colonizes bare spots created by herbicide treatment of birch. When herbicides are used it is often necessary to follow this with activities that eliminate new weeds and also encourage the establishment of target species (e.g., seeding or tilling).

4.4 FIRE

Burning is a potent method of succession management that has been used by people around the world for hundreds of years (Wright and Bailey, 1982). Two aspects of fire will be considered in this book. The first aspect is that of prescribed burning. In this chapter the effects of prescribed burning on succession in several different plant communities will be examined as will different techniques of prescribed burning. Chapter 8 will address the second aspect, natural fire management.

Prescribed burning is when fires are intentionally set to create a designed disturbance, to control colonization, or to control species performance. The use of fire is a relatively cheap succession management method, and a variety of goals can be met by prescribed burning. Prescribed burning can reduce fuels, prepare seedbeds, control unwanted or competing plant species, improve habitat for livestock and wildlife, thin forests, and control pests (Smith, 1986). Regardless of the reason for burning, fires invariably modify the successional development of plant communities by changing resource availability, species availability, plant vigour, or population structure.

Responses of individual plant species to fire are a function of various plant traits or adaptations that interact with fire. It is clear that many plant species are adapted to fire and may even require fire for regeneration (Gill, 1981; Keeley, 1981; Christensen, 1985). On the other hand, some plant species are not fire-adapted and would thus not be favoured in regenerating plant communities after fire.

Sprouting after fire from basal bud burls or lignotubers is common among shrub species in Mediterranean climates (Gill, 1981; Keeley, 1981). These structures occur below ground or have fire-resistant coverings and are protected from all but the hottest of fires. Many species of plants – including grasses, herbs, and trees with vegetative reproductive structures protected by soil, litter, or leaves – will sprout after top removal by fires if the fires are not intense and do not burn deeply. Removal of above ground parts by fires is similar to clipping in that apical dominance is eliminated and dormant buds are released.

Fire-stimulated germination of seeds is an adaptation that insures high seedling recruitment after fires. A number of woody plant species have seeds with hard seed coats that are dormant until heat-treated. The breaking of dormancy by heat coupled with site preparation by fire leads to high seedling survivorship (Christensen, 1985). Seed release may also be stimulated by fire. Closed-cone pines of North America and various woody plants in Australia retain seeds on the plant for long periods of time, releasing them only when heated. Seed retention on a plant represents an above ground seed bank that is not available until conditions for germination are optimal.

Some trees have fire-resistant bark that allows them to survive fires. Fire-resistance of this type usually develops with age. Thus, seedlings may be killed by fire but adult individuals of the same species survive. Lastly, a number of small-seeded tree species are not fire-adapted and will perish when burned but their seeds readily germinate on bare mineral soil created by hot fires.

Plants respond individualistically to fire. However, unlike clipping or herbicide application it is difficult to be species-specific in the application of fire, unless single-species stands are burned. Various characteristics of the prescribed burn such as frequency, timing, extent, and intensity can be modified to exert some control over the outcome of prescribed burning. Examples from grasslands, heathlands, and southern US pine forests will be given in the following pages to demonstrate how succession can be modified under various prescribed burning regimes. Effects of prescribed burning in other community types can be found in Kayll (1974) and Wright and Bailey (1982).

4.4.1 Woody-plant invasion of grasslands

Grassland in many parts of the world will progress to a tree- or shrub-dominated stage if fires do not occur on a regular basis (Kucera, 1981). Fires can completely kill both the seedlings and adults of woody plants or they reduce above ground stature and limit woody plant presence to sprouts. Bragg and Hulbert (1976) noted that unburned Kansas tallgrass prairie showed a 34% increase in woody plant cover from 1937 to 1969 whereas regularly burned prairie showed a low but constant woody plant cover (Figure 4.10). Among the woody plants invading the prairie were a mixture of large and small seeded species such as oaks (*Quercus* spp.), American elm (*Ulmus americana*), hackberry (*Celtis occidentalis*), and eastern red cedar (*Juniperus virginiana*). Of these species, red cedar is the most susceptible to killing by fire, but all of these species can be killed provided they are not large and do not occur in protected areas (Smith and Owensby, 1972).

Stopping woody plant invasion may involve targeting certain keystone species. These initial woody invaders or keystone species create a complex series of interactions that hasten the invasion of other woody species. For example, Petranka and McPherson (1979) found that sumach (*Rhus copallina*) colonies spreading across Oklahoma prairie inhibited grass and herb species and also facilitated the invasion of late successional tree species. Sumach may have served as

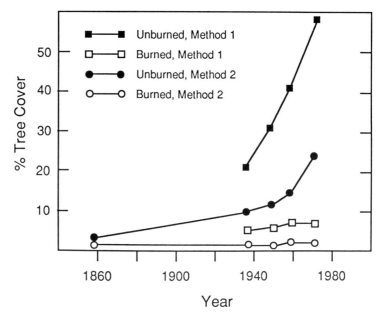

Figure 4.10 Changes in percentage tree cover of burned and unburned Kansas bluestem prairie. Method 1 involved the use of aerial photographs; method 2 involved collecting data from surveyor's records. Modified from Bragg and Hulbert (1976). Journal of Range Management (1976). Used by permission.

perching sites for birds, which then deposited tree seeds in their faeces. Burning of sumach will reduce it in stature but the number of stems may actually be increased after burning due to sprouting (Smith and Owensby, 1972). Thus, once tree invasion by some species is started, it may be necessary to continue prescribed burning at regular intervals to hold the woody plants in check.

Mesquite (*Prosopis*) is an aggressive invader of dry grasslands in the southwestern US. It also tends to facilitate the invasion of other woody species (Archer *et al.*, 1988). Late spring or early summer burning appears to be the most effective method of controlling it (Cable, 1972). Control here refers to kill or simply maintaining the mesquite individuals in a stunted condition. However, the larger mesquite individuals grow the more resistant they are to fire (Figure 4.11). Moreover, mesquite can sprout from either crown or basal stem buds depending on fire heat. The best kill of mesquite is achieved when grasses provide enough fuel to create a hot fire (Cable, 1972) as shown

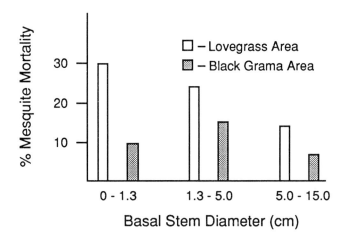

Figure 4.11 Percentage mesquite (*Prosopis juliflora*) mortality the first growing season after burning relative to stem size classes. Higher mortality in the lovegrass (*Eragrostis lehmanniana*) area was due to greater fuel accumulations. Modified from Cable (1972).

in Figure 4.11 where Lehmann lovegrass (*Eragrostis lehmanniana*) provided twice the fuel of black grama (*Bouteloua eriopoda*).

Prescribed burning in African grassland can eliminate trees or cause a shift to tree species that are fire resistant. The frequency of burning is not so important as is the simple presence of fire. Even a single fire can dramatically slow the invasion rates of trees species such as *Dichrostachys cinerea* and *Terminalia sericea* in the Transvaal Lowveld (Van Wyk, 1971). Burning when trees are under moisture stress tends to improve kill rates and also assures that fuels are readily ignited. For example, Rose Innes (1971) collected data from a number of experimental plots in west Africa and concluded that burning late in the dry season was more successful at controlling trees than either burning early in the dry season or not burning. Unburned plots in all experiments were invaded by trees whereas regular burning reduced the number of fire-tender trees and increased the number of fire-tolerant trees (Figure 4.12).

4.4.2 Fire and non-woody grassland species

Both negative and positive effects of burning have been recorded for non-woody grassland species. Where growth rates increase due to burning, usually in tallgrass or moist sites, the effect is the result of

88 *Plant and plant part removal*

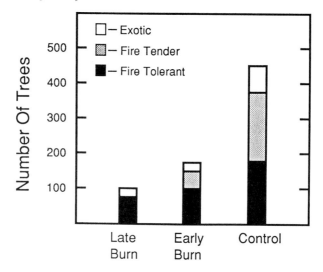

Figure 4.12 The number of exotic, fire tender, and fire tolerant trees found in permanent plots of a Nigerian savanna after 42 years of annual burning. The burning treatments were: late burn, burned late in the dry season; early burn, burned early in the dry season; and control, protected from burning. Modified from Rose Innes (1971).

litter removal, increased light availability, soil warming, and greater availability of nutrients (Kucera, 1981). Where burning decreases productivity, usually in drier areas, it is due to the indirect effects of fire which decrease moisture availability or to its direct effects which damage plants. Successional development of grasslands after fire will thus be the result of individual species responses to fire.

Plants that are actively growing will be more damaged by fire than those that are dormant (Anderson *et al.*, 1970). Grazing of grass shoots after fire will also strongly affect species persistence. If plants are burned late in the growing season before return of carbohydrates to the roots and then the new shoots growing after fire are grazed, high mortality will result (Rose innes, 1971).

Prescribed burning can favour some species over others. Smith and Owensby (1972) reviewed literature on species responses to fire in Kansas tallgrass prairie and concluded that big bluestem (*Andropogon gerardii*) was favoured by many types of burning treatments, indiangrass (*Sorghastrum nutans*) was increased by late spring burning and decreased by autumn and early spring burning, and Kentucky bluegrass (*Poa pratensis*) was decreased by all burning treatments. Van Wyk (1971) found that annual burning of South African lowveld

Table 4.6 Frequencies (number of strikes per 2000 points) of Transvaal Lowveld grasses after 15 years of different burning treatments

Species	Initial	Annual	Biennial	Triennial
Digitaria pentzii	1.56	23.63	8.30	7.55
Elyonurus argenteus	9.56	7.88	7.70	7.00
Heteropogon contortus	5.67	9.38	6.25	6.15
Hyperthalia dissoluta	17.96	3.00	6.65	3.40
Andropogon amplectens	5.45	1.00	1.70	2.50
Pogonarthria squarrosa	1.86	4.13	1.95	1.50

Modified from Van Wyk (1971).

stimulated grasses more than biennial burning or not burning. Pioneer grasses such as *Digitaria pentziii* and *Pogonarthia squarrosa* increased on annually burned plots; *Hyperthalia dissoluta*, the original dominant grass species, decreased under burning (Table 4.6). Frequent burning also tends to favour grasses over herbaceous species, perhaps due to stimulation of tillering, but there are still many questions surrounding this shift (Kucera, 1981).

4.4.3 Heathlands

Heathlands of western Europe have been subjected to management by fire during the last 200 years (Gimingham, 1970). These plant communities dominated by low-growing shrubs represent a plagioclimax plant community, one that is maintained in a semi-arrested stage of development by people. In the absence of fire, heathlands are readily invaded by trees and other species (Figure 4.13). Indeed, the cessation of active management in the form of burning, grazing, or cutting is now leading to the successional loss of many heathlands through Europe (Marrs et al., 1986).

In well-managed heathland, successional development can be observed among both dominant and subordinate species. The dominant species of many heathlands, *Calluna vulgaris*, lives for approximately 30–40 years. During that period plants go through pioneer, building, mature, and degenerate stages of development. Although biomass increases throughout the entire life cycle, productivity and nutrient content of shoots are highest during the pioneer and building stages. The grazing value of *Calluna* to both sheep and red grouse (*Lagopus lagopus scoticus*) begins to decline in the mature

90 *Plant and plant part removal*

Figure 4.13 (a) Well-managed heathland in Great Britain dominated by *Calluna vulgaris* with little invasion of other woody plant species. (b) Poorly-managed heathland invaded by bracken (*Pteridium aquilinum*), birch (*Betula pendula*) and Scots pine (*Pinus sylvestris*). Photographs courtesy of Dr R.H. Marrs.

stage (about 10 years of age), and prescribed burning is used to keep *Calluna* in the pioneer and building stages. Since *Calluna* begins to lose the ability to regenerate vegetatively after 15 years of age (Miller and Miles, 1970), fire frequency is maintained at approximately once every 15 years to conserve productive, tree-free heathland. Where fires are eliminated from *Calluna* heath, much of the *Calluna* passes into the degenerate stage. The increased openness of the community then allows invasion by birch (*Betula pendula*), Scots pine (*Pinus sylvestris*) bracken (*Pteridium aquilinum*), and other plant species (Figure 4.13).

Once *Calluna* heath becomes degenerate and is invaded by other woody-plant species, the value of fire as a management technique decreases. Large accumulations of fuel lead to intense fires that are difficult to control. Hot fires may burn into the organic layers thereby destroying the underground *Calluna* parts (Gimingham, 1970). *Calluna* regeneration must then occur from seed, a slow process. In summary, fire management of existing *Calluna* heathland is an activity that must be maintained on a strict schedule. Eliminating fire allows succession to occur; the resulting communities do not adequately reflect management goals.

Wet heathlands and bogs of northern Great Britain and Scotland do not show the same response to fire as dry heathland. Short fire rotations (<10 years) increase the dominance of graminoids and decrease *Calluna* whereas long fire rotations (20 years) increase the dominance of *Calluna*. Graminoids such as *Eriophorum vaginatum* expand vegetatively into open spots created by frequent fires (Hobbs, 1984).

4.4.4 Southern US pine forests

Prescribed burning is now used more intensively in southern pine forests than in any other community type in the US (Figure 4.14). Annual or biennial burning is a common management technique to stop or slow the encroachment of hardwood trees, shrubs, and vines in the understorey, to eliminate pine reproduction, and to eliminate litter and stimulate grass production (Vogl, 1972).

The maintenance of an open understorey in pine forests is contingent on fire either killing or pruning woody plant species in the understorey while at the same time not harming the relatively fire-resistant adult pine trees (Figure 4.15). Hodgkins (1958) noted the complexity of fire's effects in these ecosystems. Experimental burnings in Alabama loblolly (*Pinus taeda*)–short leaf (*Pinus echinata*) stands demonstrated that the first burns during stand development were

92 Plant and plant part removal

Figure 4.14 Prescribed burn in a Florida pine forest. Rapid movement of the flames is facilitated by the dense understorey of *Panicum abscissum*. Photograph courtesy of Dr P.J. Kalisz.

more effective in killing understorey trees than later burns. Understorey trees were killed more frequently when burned in August rather than January, and burns were more effective when there were large accumulations of pine needles or dead grass. However, after three years, hardwood reproduction was more dense on the burned plots (Table 4.7). Hodgkins concluded that fire is a useful method of eliminating understorey species provided the understorey trees are small, fuel is adequate, and burning is done frequently. In stands where the trees are large and fuel is sparse, partial kill may produce an understorey that is more dense than before burning.

The responses of grasses and herbaceous species to prescribed fires in pine forests are of primary importance to range managers. In general, fire tends to decrease litter and increase the cover of grasses; responses of herbaceous species are less predictable (Vogl, 1972). Protecting a pine–wiregrass savanna from fire in Georgia caused a sudden reduction in herbage yield (Lewis and Hart, 1972). This was the result of litter accumulation and of shading by gallberry (*Ilex glabra*). Prescribed burning increased the yield of grasses, but

Fire 93

Figure 4.15 (a) A long-leaf pine (*Pinus palustris*) stand in Florida 8 weeks after prescribed burning. Note the sparcity of the understorey typically dominated by palmetto (*Sabal* sp.). (b) A long-leaf pine stand 16 years after prescribed burning. Note extensive regeneration of palmetto in the understorey. Photograph courtesy of Dr P.J. Kalisz.

Table 4.7 Mean percentage cover of understorey trees, shrubs, and grasses in Alabama three years after experimental burning

Species	Treatments		
	August burn	January burn	Unburned control
Pines	0.40[a]	0.69[a]	3.39[b]
Hardwoods*	13.96[a]	11.10[a]	8.44[a]
Hardwoods†	12.10[ab]	13.08[a]	6.79[b]
Shrubs and vines	10.91[a]	18.37[b]	5.96[a]
Grasses	2.87[a]	1.64[a]	1.02[a]

* Mostly *Quercus* spp., *Carya* spp., and *Cornus florida*.
† Mostly *Diospyros virginiana*, *Rhus* spp., and *Sassafras albidum*.
Statistical comparisons are between burns done in August, January, and a control plot. Means with different letters are significantly different at the 5% level.
Modified from Hodgkins (1958). Used by permission of the Ecological Society of America.

Table 4.8 Mean crown spread (m) per 24.3 m transect of grasses, all herbaceous plants, and understorey woody plants in South Carolina forest after 20 years of repeated burning

Species group	Burning treatment					
	Unburned control	Periodic winter	Periodic summer	Annual winter	Biennial summer	Annual summer
Grasses	1.0[a]	1.1[a]	2.0[ab]	2.7[b]	2.2[ab]	1.3[ab]
Herbaceous plants	1.0[a]	1.5[a]	2.3[a]	6.1[b]	5.8[b]	3.4[ab]
Woody plants	11.0[c]	12.1[c]	11.3[c]	5.9[b]	3.7[b]	0.5[a]

Means within a species group with different letters are significantly different at the 5% level.
Modified from Lewis and Harshbarger (1976). *Journal of Range Management* (1976).
Used by permission.

regeneration of gallberry produced more sprouts on the burned plots as compared to unburned plots. Again, the results suggest that if burning is to be used as a method of controlling understorey growth and improving grazing, then it must occur at frequent intervals (Figure 4.15).

Lewis and Harshbarger (1976) summarized the results of a 20-year study in which they monitored the development of understorey species in South Carolina loblolly pine forests under different burning regimens. They found that all treatments – with the exception of annual summer burning – increased the ground cover of plants as compared to controls. Grasses and shrubs were favoured by different burning regimens (Table 4.8). Grasses were favoured by annual winter and biennial summer burns, but shrubs were not. Low-growing shrubs regenerating as sprouts were favoured by periodic summer burning or winter burning. In terms of wildlife and grazing, Lewis and Harshbarger concluded that annual or periodic winter burning was the best management approach because it favoured grasses for cattle, forbs for quail, and shrubs for deer and turkey.

4.4.5 Prescribed burning techniques

Although burning prescriptions are now available for a wide variety of plant community types, Wright and Bailey (1982) maintained that prescribed burning is still 'both a science and an art'. A full consideration of burning prescriptions will not be attempted here, but mention of important factors controlling fire behaviour and descriptions of established firing techniques will demonstrate what is involved if prescribed burning is chosen as a management technique For a complete treatment of prescribed burning in different community types readers are referred to Wright and Bailey (1982) and Mobley *et al.* (1973).

Resource managers considering prescribed burning should consult with individuals who have expertise before attempting a burn. Using fire as a succession management technique requires much pre-burn planning, coordination of personnel, communication with landowners and local fire prevention groups, a knowledge of weather and fuel conditions, and proper equipment. Smoke generated by prescribed fires is also a factor that must be considered in areas of high human populations.

(a) Fuel

Characteristics of fuel that are most important in prescribed burning are the fuel load or mass per unit area, fuel volatility, and fuel moisture

96 *Plant and plant part removal*

content. Minimum fuel loads are necessary to carry a fire and to achieve kill of shrubs and trees. In grasslands at least 500 kg ha^{-1} of fine fuel are necessary, although this can be less if shrubs in the grassland are closely spaced (>20% cover) and the shrubs are carrying some dead material (Wright and Bailey, 1982). The presence of oils, waxes, or fats in fuels such as in conifers, shrubs of Mediterranean ecosystems, and Australian eucalypts, makes for high volatility. On the other hand, grasses and most hardwood trees produce fuels of low volatility. Highly volatile fuels increase the fire intensity potential and also produce firebrands, glowing particles of fuel that detach during burning. These firebrands increase the danger of spot fires outside the boundaries of the prescribed burn. Fine fuels will not burn when the moisture content is above 30–40%; the preferred moisture range for burning fine fuels is 7–20% (Wright and Bailey 1982; Mobley *et al.*, 1973).

(b) Weather

Fuel moisture is directly related to relative humidity (RH). The optimum RH for burning is 25–40%. Above 40% RH fires do not carry well, and below 25% RH spot fires from firebrands are a problem (Wright and Bailey, 1982). Temperature influences both the danger of spot fires from firebrands and the ignition of fuels. Above 15 °C the danger of unwanted spot fires increases exponentially (Bunting and Wright, 1974); below 0 °C burning tends to be sluggish (Wright and Bailey, 1982). Some wind is necessary to carry prescribed fires because this is the primary way in which oxygen is replenished. Optimum wind speeds for burning vary with community type. Topkill of standing trees in rangeland requires winds in the range 13–24 km h^{-1} but not above 32 km h^{-1} (Wright and Bailey, 1982).

(c) Firelines

Firelines are used to stop fire movement at predetermined boundaries (Figure 4.16). Firelines such as lakes, roads, or ploughed fields may be in place before prescribed burning and should be used whenever possible to minimize costs and disturbance. More commonly, however, firelines must be constructed by ploughing and burning, bulldozing and burning, mowing and raking, or by wetting with sprinklers. Regardless of fireline construction methods, the goals are the same: to eliminate burnable fuel and contain the fire within the fireline boundaries. Fireline width varies with fuel type and community type. Firelines can be 2–6 m in grasslands. They must be at least 75 m with moderately volatile fuels and at least 150 m with

Figure 4.16 A ploughed fireline that successfully stopped a prescribed burn in a Florida longleaf pine (*Pinus palustris*) forest.

highly volatile fuels such as in chaparral communities (Wright and Bailey, 1982).

(d) Burning techniques

A large variety of burning techniques have been tested in different plant communities (Figure 4.17). The utility of a specific burning technique will depend on fuel characteristics and the danger of unwanted fire. Headfires are fires that move with the wind; backfires move into the wind. Headfires will be most effective at killing trees and shrubs when fuel accumulations are low; backfires are cooler and easier to control (Wright and Bailey, 1982). Strip headfires and flank fires provide somewhat more control than headfires; centre ignition and spot ignition techniques generate intense fires used primarily for the burning of slash (Wright and Bailey, 1982).

4.5 CABLING

Cabling is a technique used to modify the stature of woody plants. It involves linking two bulldozers with a heavy cable or chain that is

98 Plant and plant part removal

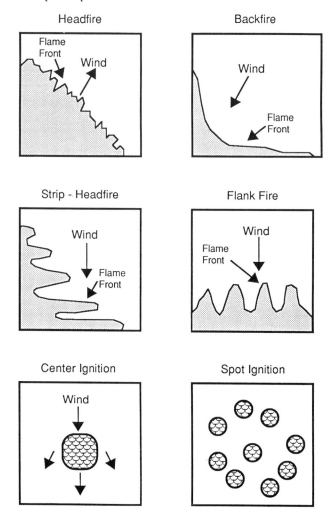

Figure 4.17 Various ignition patterns used in prescribed burning. Scalloped areas indicate flames and flame fronts relative to wind direction. Darkened borders indicate firelines. Reprinted by permission of John Wiley and Sons Inc., all rights reserved.

then pulled across the landscape. Cabling was originally developed as a method to improve range for livestock in the southwestern US because the woody plants are broken off or tipped over and grasses remain undamaged. However, the value of cabling in improving range for livestock is questionable; in terms of woody plant species, some show population increases after cabling and some show population decreases. Research aimed at following community development after cabling indicates that it can modify successional pathways at least over a short-term period of several decades.

Responses of cable-damaged plants will depend on the success of vegetative reproduction or establishment from seed. For example, Schmutz et al. (1959) found that cabling reduced the size of shrubs but significantly increased shrub density. The large increase in shrub density was due mostly to jumping cholla (*Opuntia fulgida*). This shrub was broken apart by cabling, but the plant segments sprouted and then rooted at the joints, thus increasing population size. Pencil cholla (*Opuntia arbuscula*), on the other hand, lacking successful regeneration from fragments was decreased. Grasses and forbs did not respond positively to cabling (Schmutz et al., 1959). Rippel et al. (1983) found that cabling of pinyon–juniper communities in New Mexico produced a long-lasting influence on community development. Alligator juniper (*Juniperus deppeana*) was decreased by cabling whereas skunkbush (*Rhus trilobata*) was increased (Table 4.9). Other shrubs and trees showed no significant differences in density between cabled plots and control plots. Valuable forage grasses actually showed less coverage 20 years after cabling, thus not supporting the use of this technique to improve range for livestock.

Table 4.9 Mean density (no. ha^{-1}) for tree and shrub species and mean percentage basal cover of grasses and forbs 20 years after cabling of a New Mexico pinyon–juniper community

Species	Treatment	
	Cabled	Control
Trees		
Juniperus monosperma	265[a]	312[a]
Juniperus deppeana	22[a]	57[b]
Pinus edulis	242[a]	305[a]
Shrubs		
Rhus trilobata	435[a]	262[b]
Opuntia inbricata	35[a]	85[a]
Berberis haematocarpa	15[a]	13[a]
All grasses	20[a]	26[b]
All forbs	9[a]	14[b]

Means with different letters are significantly different at the 10% level. Modified from Rippel et al. (1983). Used by permission of the Journal of Range Management.

Overall, cabling does not appear to positively influence the establishment or growth of forage grasses. However, cabling can strongly modify the population structure of shrubs and woody plants. Those species that can regenerate quickly from seeds in the soil or from vegetative reproduction will dominate communities for long periods after management (Schott and Pieper, 1987).

SUMMARY

Above ground plant or plant-part removal in succession management is achieved by cutting, herbicides, burning, or cabling. Such plant removal may be temporary if tops only are killed or it may be permanent if plants are root killed. Sprouting patterns of top-killed plants and the responses of non-target species to release from competition are important determinants of succession following management. The creation of bare soil is also important.

Cutting immediately reduces the above ground stature of plants and eliminates apical dominance. Those species that cannot sprout from structures at or below the soil surface are eliminated. Repeated cutting can change species dominance in grasslands, can reduce populations of unwanted species in heathland, and can stimulate the regeneration of non-target species from seed. Forests regenerating after clearcutting may become dominated by shrubs released from competition or by tree species capable of stump sprouting. Variations in the timing of cutting can modify the responses of target and non-target species to cutting.

Depending on the method of application, the timing of application and the herbicide type, herbicide use will achieve either top-kill or root-kill. Top-killed plants that can sprout respond as if cut. Successful herbicide use in succession management depends on selectivity in herbicide application and herbicide action. Screening trials are recommended to determine the effects on target and non-target species before widespread spraying occurs. The most selective herbicide application methods are those applying herbicides directly to cuts or gashes on the target organism. Basal herbicide application to trees in rights-of-way can lead to low-growing vegetation that resists tree invasion and satisfies management goals. Spraying of woody plants in rangelands may or may not lead to long-term production increases of grasses and herbaceous species.

Prescribed burning is when fires are intentionally set and then controlled in extent, frequency, and intensity. Burning may be used to eliminate or stimulate target species. In grasslands, fire will stimulate the production of some grass species but will kill or stunt woody plant species. The resistance of woody plants to fire increases with plant size. Depending on the frequency and timing of burning, fire will encourage or discourage grasses and shrubs in the understorey of southern US pine forests. In short, plants respond individualistically to fire, and the response will be determined by the presence of certain fire adaptations. If fires do not burn deeply into the soil, plants that reproduce vegetatively respond to fire as if cut. Successfully carrying out a prescribed burn requires proper construction of firelines as well as knowledge of fuel and weather conditions.

Cabling breaks or pushes over tall woody plants but leaves low-growing plants unharmed. Woody plants that can reproduce by fragmentation or from buried seeds are stimulated by cabling whereas others may decrease. Production of grasses and herbaceous species is not generally increased by cabling, but dominance of shrubs and trees can be modified.

– 5

Methods of managing succession: changing resource availability

5.1 INTRODUCTION

Clipping, burning, and spraying do indeed change resource (light, water, and nutrient) availability for surviving plants during succession management and thus can be considered designed disturbances. For example, burning of plant tissue may volatilize its contained nitrogen, which is then no longer available for on-site plant use. Clipping of canopy plants may change the light regimen of understorey plants. These effects, however, are indirect. Resource availability can also be directly manipulated to modify plant establishment, growth, or competition (i.e., controlled colonization or controlled species performance). Included in this chapter are techniques that change nutrient availability (fertilization or nutrient exhaustion) and water availability (irrigation or water level change).

5.2 SOIL NUTRIENTS

Plant species express differences in nutrient requirements and differences in their growth responses to nutrient availability. This allows community development to be modified by changing the nutrient environment. Chapin (1983) characterized a spectrum of nutrient use for plant species adapted to different environments. One end of the spectrum included early successional species that are annuals and have rapid growth rates, high rates of nutrient absorption, and dramatic growth responses to fertilization. On the other end of the spectrum are slow-growing plant species adapted to infertile environments. They show low nutrient absorption and less dramatic growth responses to fertilization (Chapin, 1983). This spectrum suggests that soil nutrient availability is an important determinant of plant species composition at a single point in time and also an important determinant of plant species turnover during succession.

Because many successional pathways begin with annual species and then progress to perennial species, there may be gradual changes in the

104 *Changing resource availability*

responsiveness of a community to nutrient additions or deletions. The growth of small-seeded annual plants is directly tied to soil nutrient availability. The period of maximum annual plant growth coincides with maximum nutrient uptake (Williams, 1955). The growth of perennial plants, on the other hand, is partially dependent on stored nutrient in roots, stems, or large seeds and is less directly tied to soil nutrient availability. Nutrient uptake in perennials often coincides with the time when nutrients are most available (Chapin, 1983). Thus it is likely that early stages of succession are more responsive to fertilization than are later stages (Mellinger and McNaughton, 1975); it is not surprising that a majority of the research where succession is modified by nutrient additions or deletions deals with early successional communities.

There are a number of reasons why resource managers might be concerned with the effects of nutrient availability on succession. The addition of various nutrients can modify competitive relationships. Nutrient regimens can be avoided that would eliminate less competitive species; nutrient regimens can be created to encourage other species. In addition, eutrophication is a common environmental problem and resource managers should be able to predict how this will affect succession on a parcel of land.

5.2.1 Old fields

Although old fields in the US are a valuable resource for the generation of data and theories on succession, there are few applied research projects now in progress where the effects of long-term fertilization are being monitored. Most of the studies completed up to now are of a short-term nature. In them, the reason for establishing fertilization treatments was not to manipulate succession but rather to understand why succession occurs. Nevertheless, the short-term studies that are available can provide information on what trends might be expected if old fields are fertilized or if they are used as sites for the disposal of nutrient-rich wastes.

In old fields, nitrogen is the most limiting nutrient. Only when nitrogen is readily available do other nutrients such as phosphorus (Mellinger and McNaughton, 1975) or magnesium (Tilman, 1984) become important. When old fields are fertilized it is not uncommon to find that some species respond positively and others respond negatively (Mellinger and McNaughton, 1975; Bakelaar and Odum, 1978). Both the nutrient requirements of individual species and changing competitive interactions may be responsible. For example, Bakelaar and Odum (1978) found that the perennial herb *Solidago*

Table 5.1 Mean above ground net production by species ($g\,m^{-2}\,yr^{-1}$) of an 8-year-old Georgia successional field during 1974 when fertilized in late winter 1973 and early spring 1974

Species	Control	Fertilized	Significant difference
Solidago canadensis	374	923	yes
Aster pilosus	121	12	no
Lactuca scariola	70	31	no
Sorghum halepense	69	179	no
Andropogon virginicus	55	31	no
Bromus secalinus	39	217	yes
Lolium multiflorum	36	91	yes
Cyperus spp.	23	12	no
Allium vineale	17	7	no
Bromus japonicus	17	58	no
Daucus carota	14	3	no
Vicia angustifolia	14	15	no
Heterotheca latifolia	13	1	no
Poa spp.	12	30	no
Oxalis stricta	12	8	no
Gnaphalium obtusifolium	11	2	no
Campsis radicans	9	2	no
Cassia tora	5	1	no
Rumex crispus	5	3	no
Total	946	1672	yes

Fertilization rates were $560\,kg\,ha^{-1}$ and $448\,kg\,ha^{-1}$ of 6-12-12 fertilizer. $n = 2$.
From Bakelaar and Odum (1978). Used by permission of the Ecological Society of America.

canadensis was greatly stimulated in a Georgia old field by fertilization but the production of *Aster pilosus* was curtailed presumably following increased competition from *Solidago* (Table 5.1).

The question of old field responsiveness to fertilization as a function of field age has not been answered and existing data are contradictory. Mellinger and McNaughton (1975) compared the responses of 6- and 17-year-old fields in New York to fertilization with $560\,kg\,ha^{-1}$ of 10-10-10 fertilizer. Only the younger field showed an increase in production. In contrast, Carson and Barrett (1988) fertilized young (1-year-old) and old (3-year-old) experimental plots in Ohio successional fields with either fertilizer or sewage sludge (Figure 5.1).

106 *Changing resource availability*

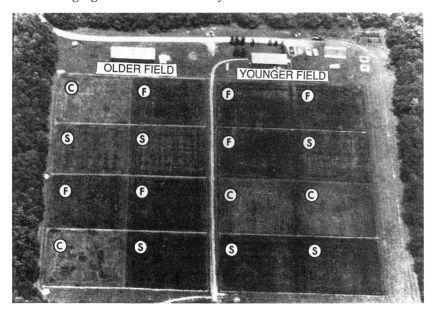

Figure 5.1 Old field experimental plots in Ohio that received either artificial fertilizer (F) or sewage sludge (S). The nutrient input was ca. 540 kg ha^{-1} N and 170 kg ha^{-1} P. The younger field was 1 year old and the older field was 3 years old. Photograph courtesy of Dr W.P. Carson.

In the young field net primary production during the first year (1978) was significantly higher in plots with either fertilizer or sludge but this effect did not persist to the last year of the study (Table 5.2). Total net primary production in the older field was also significantly higher in enriched plots during the last year of the study. Clearly, more research is needed before generalizations about age of an old field and its response to fertilization can be made. However, the research of Carson and Barrett (1988) suggests that fertilization (Table 5.2) tends to increase the importance of annual species characteristic of early successional stages such as *Ambrosia artemisiifolia*, *Polygonum persicaria*, and *Setaria faberi*. If these results have wide applicability, then fertilization may indeed be a method of reversing succession in old fields.

5.2.2 Heathlands

Artificial fertilization of heathlands leads to a dramatic shift in species composition. Extant heathland species typically do not show positive growth responses to high nitrogen availability (Heil and Diemont,

Table 5.2 Mean above ground net production (g m^{-2} yr^{-1}) of plant species in young (1 year) and older (4 year) Ohio old fields 2 years after fertilization with 1570 kg ha^{-1} of 34-11-0 dry fertilizer (F) or 8960 kg ha^{-1} of 6-2-0 dry municipal sludge (S). The young field was fertilized in autumn 1977; the older field, in May 1978

Species (young field)	1978			1980		
	F $n = 3$	S $n = 3$	C $n = 2$	F $n = 3$	S $n = 3$	C $n = 2$
Ambrosia artemisiifolia	877^a	456^b	151^c	262^a	11^a	3^a
A. trifida	2^a	0^a	0^a	96^a	0^a	0^a
Aster pilosus	0^a	0^a	0^a	0^a	24^a	751^b
Chenopodium album	62^a	5^a	6^a	85^a	19^{ab}	0^b
Cirsium arvense	18^a	52^a	89^a	262^a	374^a	61^b
Daucus carota	0^a	0^a	0^a	0^a	0^a	2^a
Erigeron annus	0^a	1^a	1^a	0^a	0^a	368^b
Melilotus officinalis	5^a	2^a	26^b	0^a	0^a	0^a
Polygonum persicaria	12^a	15^a	1^a	349^a	314^a	1^b
Potentilla norvegica	0^a	0^a	0^a	0^a	0^a	0^a
Setaria faberi	43^a	161^a	11^b	278^a	165^a	40^b
Solidago canadensis	3^a	0^a	0^a	16^a	66^a	55^a
Trifolium pratense	14^a	38^a	110^a	0^a	0^a	8^a
Triticum aestivum	536^a	326^a	364^a	0^a	1^a	0^a
Total	1573^a	1048^b	733^c	1582^b	1171^a	1597^b
Species (older field)						
Ambrosia artemisiifolia	0^a	0^a	0^a	392^a	169^b	0^b
Barbarea vulgaris	0^a	0^a	0^a	161^a	181^a	0^b
Dactylis glomerata	105^a	48^a	36^a	9^a	0^a	17^a
Festuca elatior	350^a	177^b	125^c	215^a	91^a	246^a
Phalaris arundinacea	270^a	200^a	128^a	0^a	0^a	0^a
Poa pratensis	120^a	100^a	75^b	11^a	107^a	126^a
Polygonum persicaria	0^a	0^a	0^a	0^a	171^a	0^a
Setaria faberii	11^a	13^a	0^a	407^a	260^a	0^b
Solidago canadensis	120^a	112^a	21^a	217^a	772^a	140^a
Vernonia altissima	0^a	0^a	0^a	0^a	0^a	66^a
Total	1124^a	747^b	465^c	1503^a	1971^b	648^c

Treatment means for each year with different letters are significantly ($P < 0.05$) different.

From Carson and Barrett (1988). Used by permission of the Ecological Society of America.

108 Changing resource availability

1983) or high nitrogen and phosphorus availability (Specht et al., 1977). In fact, many heathland species are adapted to low nutrient environments and when fertilized their importance tends to decrease as shown in Table 5.3 where species such as *Dillwynia peduncularis*, *Homoranthus virgatus*, and *Leucopogon ericoides* growing in Australian heathland decreased in importance after fertilization. Similar results were noted for *Calluna vulgaris* growing in fertilized Swedish heathland (Persson, 1981). On the other hand, various grass species growing in or around heathlands respond positively to fertilization, especially nitrogen. This may rapidly change heathlands to grass-dominated communites. Heil and Diemont (1983) described the conversion of British heathland into communities dominated by the grass *Festuca ovina* when nitrogen was repeatedly applied. Specht et al. (1977) noted that the Australian grass *Themeda australis* filled gaps left by heathland species (Table 5.3). Fertilization of heathlands is clearly a successful method of succession management primarily when the management goal is to convert heathlands into pasture. Unfortunately, this is typically not the management goal but rather the opposite: convert grassland into heathland.

Table 5.3 Mean density (plants m^{-2}) and biomass (g m^{-2}) of selected Australian heath plants measured during spring 1976. Fertilizer treatments were applied in 1968 (95 kg ha^{-1} P, 63 kg ha^{-1} N, 138 kg ha^{-1} S and 184 kg ha^{-1} Ca)

Observation	Control	Fertilized	Significant difference ($P < 0.05$)
Density			
Dillwynia peduncularis	0.3	0.2	yes
Homoranthus virgatus	0.5	0.1	yes
Leucopogon ericoides	0.4	0.3	yes
Biomass			
Homoranthus virgatus	618	95	yes
Caustis recurvata	65	7	yes
Coleocarya gracilis	139	49	yes
Themeda australis*	11	29	yes

* Savannah grass species.
From Specht et al. (1977).

5.2.3 Disturbed soils

Revegetated land surfaces provide excellent opportunities for assessing the effects of nutrient availability on plant community development because of the fact that the starting nutrient capital in the soil is usually very low (Bloomfield *et al.*, 1982). Specific amounts of nutrients may be added to disturbed soils by artificial fertilizers or nitrogen may be added by the seeding of nitrogen-fixing legumes. Those studies in which succession on disturbed soil was modified by artificial fertilization will be considered first. A subsequent chapter will consider the effects of legumes because in all situations where legumes are artificially planted the competitive effects of the legume interact with nitrogen fertilization.

As in old fields, nitrogen is the primary limiting element on highly disturbed soils. Attempts to demonstrate that other elements such as phosphorus are important in the structure of communities have not been conclusive. Applying nutrients other than nitrogen in fertilizers may be a waste of time and resources.

On highly disturbed soils that are revegetated, the dominance of grasses is promoted by applying high levels of nitrogen-containing fertilizers. For example, Doerr and Redente (1983) initiated a study of plant community development on disturbed soils in Colorado (Figure 5.2). Grass biomass was significantly higher on fertilized plots in the first 2 years of the study whereas forb biomass was higher on unfertilized plots in the last year (Figure 5.2). Greater forb biomass on unfertilized plots was attributed to less competition from grasses. While the grass response to fertilization was dramatic, it appeared to be relatively short lived; by the third year after fertilization, grass production was not significantly different from unfertilized controls.

Shifting competition in favour of grasses may also be done by extending the duration of fertilization. Depuit and Coenenberg (1979) applied fertilizers at increasingly longer durations to Montana stripmine soils. With increasing duration of fertilization, perennial grass biomass increased and legume biomass decreased (Figure 5.3). Specifically, introduced perennial grasses such as crested wheatgrass (*Agropyron cristatum*) and smooth bromegrass (*Bromus inermis*) showed the greatest positive responses to fertilization, and native perennial grasses such as *Stipa viridula* and *Oryzopsis hymenoides* declined in importance. Cicer milkvetch (*Astragalus cicer*), an introduced legume was responsible for the majority of legume biomass decrease under heavy fertilization. Thus, high rates of nitrogen fertilization may produce a productive sward of a few grass species, but other less competitive species will be eliminated.

110 *Changing resource availability*

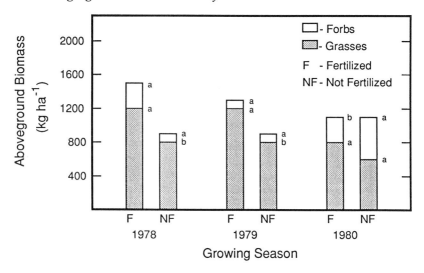

Figure 5.2 Mean biomass of grasses and forbs after fertilization of disturbed soils in Colorado. Fertilizers (112 kg ha^{-1} N and 90 kg ha^{-1} P) were applied once in 1977 and biomass measurements were taken in 1978–1980. Means with different letters within a year are significantly different ($P < 0.10$). From Doerr and Redente (1983). By permission of Elsevier Science Publishers.

If the management goal is to increase or maintain plant species diversity, then high rates of nitrogen fertilization are not recommended. Instead, low rates of fertilization (< 40 kg ha^{-1}) may increase species diversity by allowing a large number of species to establish and survive on nutrient-poor soils without permitting a few grass species to assume dominance (Doerr and Redente, 1983; Davis *et al.*, 1985).

5.2.4 Grasslands

Fertilization studies on grasslands in Great Britain indicate many of the same results as on disturbed soils. Heavy fertilization with nitrogen leads to the dominance of a few grass species such as *Alopecurus pratensis, Festuca rubra*, and *Agrostis stolonifera* while forbs are eliminated (Willis, 1963; Rorison, 1971). Furthermore, some grass species such as *Anthoxanthum odoratum* appear to be favoured by high phosphorus levels (Rorison, 1971).

In the plains region of the US fertilization causes a shift among cool and warm season grass species and nitrogen is a critical limiting element. When grasslands are fertilized there is increased importance

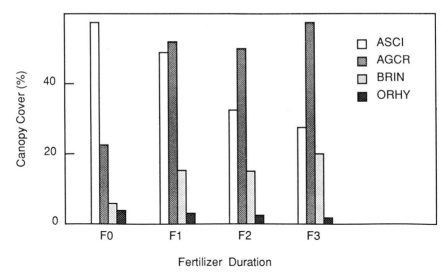

Figure 5.3 Canopy cover percentage of different plant species growing on disturbed soils in Montana under various fertilization regimens. Approximately 77 kg ha^{-1} N and 41 kg ha^{-1} P were applied in the first fertilization and 77 kg ha^{-1} N and 11 kg ha^{-1} P were applied in subsequent fertilizations. F0 = no fertilizer; F1, fertilized in 1975 only; F2, fertilized in 1975 and 1976; F3, fertilized in 1975, 1976, and 1977. AGCR, *Agropyron cristatum*; ASCI, *Astragalus cicer*; BRIN, Bromus inermis; ORHY, *Oryzopsis hymenoides*. From Depuit and Coenenberg (1979). Used by permission.

of cool season grass species (Owensby *et al.*, 1970) or weedy species (Huffine and Elder, 1960). This shift can be avoided by application of fertilizers later in the growing season. For example, Rehm *et al.* (1972) reported the results of a 4-year fertilization study in Nebraska prairie dominated by warm season grasses (Table 5.4). Production of these grasses was increased by N or N + P fertilization, but the composition of the community was unchanged. The lack of species shift in this study was attributed to the fact that increased nutrient availability coincided with the period of maximum warm-season grass growth. It is unknown, however, if continued late fertilization would eventually usurp the cool-season species.

5.3 NUTRIENT EXHAUSTION

While it is possible to change plant community development trends with the addition of nutrients, it may also be possible to modify plant

112 Changing resource availability

Table 5.4 Effects of fertilization (kg ha^{-1}) on the percentage of total biomass contributed by each species in a Nebraska prairie

Treatments			% of community biomass by species				
N	P	K	Andropogon gerardii	Bouteloua curtipendula	Panicum virgatum	Cool season species	Weeds
—	—	—	63[a]	29[a]	0.4	0.0	0.0
45	—	—	63[a]	37[a]	0.0	0.0	0.4
90	—	—	48[a]	49[a]	0.4	0.4	0.8
90	22	—	66[a]	31[a]	2.1	0.4	0.4
90	45	—	69[a]	33[a]	0.0	0.4	0.4
90	22	22	65[a]	33[a]	1.7	1.3	0.0

Fertilizers were applied each year after 15 May. Treatments began in 1967 and measurements were taken in 1971. Means in any one column with the same letter are not significantly ($P < 0.05$) different
From Rehm et al. (1972). Journal of Range Management (1972). Used by permission.

communities by the removal of nutrients (i.e., nutrient exhaustion). The impetus for using nutrient exhaustion as a management technique is linked to the observation that soil nutrient accumulation during the course of ecological succession is a primary factor driving species turnover (Green, 1972). If soil nutrient availability can be decreased, then succession can be arrested or set back to an earlier stage of development. Unfortunately there are few data on the use of fertility reduction as a means of managing succession.

The accumulation of soil nutrients during succession has been a central focus of succession research for many years. In general, the accumulation of nutrients in an ecosystem will depend on the balance of inputs and outputs. Biologically important elements such as nitrogen and phosphorus accumulate largely in soil organic matter or in plant biomass during succession. Nutrient inputs occur through rainfall, dryfall, weathering, faeces and urine, fertilization, and nitrogen fixation. Outputs occur via leaching, surface runoff, volatilization, or the removal of soil or plant biomass. In general, the retention of biologically important nutrients is dependent on the activity of plants. The higher the net increment of biomass on a yearly basis the lower will be the loss of nutrients from an ecosystem

(Vitousek and Reiners, 1975). Of critical importance to succession managers is the fact that early stages of succession are typically characterized by net nutrient retention and will not lose large quantities of nutrients by leaching unless severely disturbed (Likens et al., 1970). Methods that successfully decrease nutrient availability in ecosystems must equal and exceed the natural processes that add nutrients to soils without destroying the integrity of the plant community. At present, the most common methods of accomplishing this goal include grazing, burning, and mowing (Green, 1972), although other methods have been tested (Marrs, 1985b).

Cropping and subsequent removal of biomass can be used to reduce soil nutrient availability. This technique is implemented to encourage species characteristic of oligotrophic systems such as heathlands on soils that previously were farmed and received high inputs of artificial fertilizers. If cropping is used to reduce soil fertility, the most rapid reductions will occur if the greatest amounts of biomass are removed. Table 5.5 indicates the levels of nutrients that were removed when cereal rye was cropped on an area targeted for heathland restoration (Marrs, 1985b). Over the course of 3 years of rye cropping, removal of rye grain caused eight times more phosphorus and three times more potassium to be lost from the system than was gained from various inputs but no net loss of nitrogen was measured (Table 5.5). If both the grain and the rye biomass were removed even greater nutrient reduction could be achieved (Table 5.5) and still more nutrients,

Table 5.5 Estimated inputs and removals of nutrients (kg ha^{-1}) as a result of 1 year of cereal rye cropping in an English field

Sources	*Elements*		
	N	P	K
Inputs			
Rainfall	16.8	0.1	3.2
Cereal seed	2.1	0.6	1.6
Total	19.0	0.7	4.8
Removals			
Grain	18.0	5.0	14.0
Total above ground biomass	35.0	10.0	61.0

From Marrs (1985b). Used by permission of Elsevier Applied Science Publishers.

especially nitrogen, could be removed if rye stubble was burned. Although this nutrient reduction was apparently not successful in restoring the land to its original heathland community, the community was changed in composition to one dominated by unproductive grass species with low nutrient requirements.

Shrub or tree invasion can enrich the upper soil layers in nitrogen and phosphorus. This enrichment may in turn allow the invasion of plant species with high nutrient requirements. Hodgkin (1984) documented such a trend with the invasion of hawthorn (*Crataegus monogyna*) onto English dunes. As hawthorn died, the gaps were filled by weedy species with high nutrient requirements and the desirable dune species were displaced. In this system as well the only approach to reverse the buildup in fertility is to remove the shrubs and also to remove as much understorey biomass as possible. When this activity is repeated numerous times, soil nutrient levels may approach what is typical of oligotrophic dune communities. The rate of successional reversal will depend on the propagule supply that is left after biomass removal efforts.

Clearly, fertility reduction in succession management is not well studied. While it is relatively simple to achieve soil fertility reduction, the direct effects of activities that must be used to reduce soil fertility may have a greater influence on succession than the withdrawal of nutrients. As such, more research is needed to assess the long-term effects of fertility reduction as well as to develop new methods of achieving fertility reduction.

5.4 WATER

Because water is the most limiting molecule for plant growth in many terrestrial communities, it is not surprising to find that manipulation of water availability can change successional pathways. Nearly every aspect of plant life history is dependent on various degrees of water availability. Differences in species distributions relative to water gradients are tied to differences in physiological and anatomical features that control water acquisition or loss, desiccation, and inundation tolerances. The water supply of plants can be changed in a number of ways but only a few such techniques have relevance in succession management. This section will consider, first, the effects of irrigation on various plant communities and, second the effects of drawdown and flooding on plant communities that occur in wetlands or at lake margins. Irrigation is a powerful but expensive method of controlling species performance. Water-level manipulation functions as a designed disturbance as well as a method for controlling colonization and species performance.

5.4.1 Irrigation of grassland

Irrigation is commonly used in agriculture, especially in arid regions of the US, to supplement rainfall so that productivity can be maximized. It is also a technique that can be used to change the species composition of rangeland. In general, warm season grasses are favoured more by irrigation than cool season species. For example, Owensby *et al.* (1970) carried out a 4-year study of Kansas bluestem range where irrigation and nitrogen fertilization were done singly and in combination during each year. The warm season species big bluestem (*Andropogon gerardii*) showed significantly increased basal cover with the addition of water and water + nitrogen, whereas Indiangrass (*Sorghastrum nutans*) responded positively only to irrigation (Table 5.6). Kentucky bluegrass (*Poa pratensis*) and tall dropseed (*Sporobolus asper*), cool season grasses that increase under grazing or fertilization, did not respond to irrigation either alone or in combination with nitrogen. Moisture addition, apparently, favoured

Table 5.6 Mean percentage cover of plant species from a Kansas prairie under different nitrogen and irrigation treatments

Species	Treatments				
	Nitrogen	Moisture	Moisture + nitrogen	Control	$LSD_{0.05}$*
Decreasers					
Andropogon gerardii	35.5	39.5	41.2	34.3	4.8
Sorghastrum nutans	16.8	23.9	20.6	20.1	3.2
Increasers					
Sporobolus asper	3.3	1.6	2.2	2.1	1.0
Poa pratensis	11.4	5.9	9.1	7.4	2.2
Bouteloua curtipendula	3.9	2.2	2.9	5.1	1.6

Fertilization with N ($56 \, kg \, ha^{-1}$) was done each year from 1965 to 1968. Irrigation was carried out on a regular basis throughout the study
* Least significant difference among treatment means ($P < 0.05$).
From Owensby *et al.* (1970). *Journal of Range Management* (1970). Used by permission.

116 Changing resource availability

the warm season species to the competitive suppression of cool season species.

5.4.2 Irrigation of disturbed soils

Irrigation is also used in arid areas when disturbed soils are revegetated. Moisture additions tend to speed both plant establishment and successional change, but the long-term effects of such treatments are largely unknown. When an irrigation treatment is applied, the effects are short lived as indicated by the data on Colorado coal mine spoils in Figure 5.4. In general, grasses tend to be more responsive to irrigation than forbs. Considering the short-lived effects and the high cost of irrigation, it is not recommended as general practice at least in arid regions (Doerr and Redente, 1983).

5.4.3 Irrigation of Swedish pine-heath

Irrigation can interact in a complex manner with fertilization. The responses of plant species are highly individualistic. For example, Persson (1981), working in a Swedish pine-heath ecosystem, began

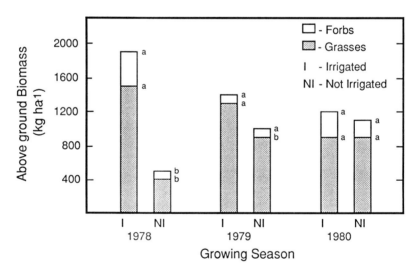

Figure 5.4 Mean biomass of grasses and forbs growing on disturbed soils in Colorado after irrigation (2.5 cm week^{-1}). Plots were irrigated during the 1977 and 1978 growing seasons but not during the 1979 season. Means with different letters within a year are significantly different ($P < 0.10$). From Doerr and Redente (1983). By permission of Elsevier Science Publishers.

irrigation and fertilizer treatments in 1974 and continued them through 1980 during which time water was sprayed on plots throughout the growing season at the rate of $3 \, l \, m^{-2} \, day^{-1}$. Nitrogen fertilization totalled 80 kg ha^{-1} yr^{-1}. Irrigation alone did not cause significant changes in plant cover during the study period. When plots were irrigated and fertilized, however, dramatic changes in species cover were recorded. *Calluna vulgaris* showed a gradual decline; *Vaccinium vitis-idaea* increased during the middle stages of the study and then decreased toward the end; *Chamaenerion angustifolium* showed a strong increase as compared to controls. Lichens nearly disappeared from the irrigated/fertilized plots and the moss *Pohlia nutans* expanded and filled the gaps left by lichens. Presumably, increased moisture availability makes fertilizers more available to plants and individualistic responses are more readily expressed.

Because of the difficulties associated with applying irrigation treatments and because of the short-lived effects, it is not likely that this method of managing succession will become widespread in the future.

5.4.4 Water-level manipulation

In contrast to irrigation, manipulation of the water level in wetlands, ponds, lakes, or reservoirs is a powerful method of controlling colonization and species performance among sediment colonizing, floating aquatic, and emergent plant species. Drawdown or flooding are only possible if the amounts of water entering or leaving a body of water are controllable. This management technique is typically limited to man-made impoundments.

Drawdown has long been used in the management of man-made impoundments as a method of controlling weedy plant species (Cooke *et al.*, 1986). By lowering the water level these plants are exposed and may then be killed by desiccation, heat or freezing. Early research on the drawdown of large reservoirs in the US, however, indicated species-specific effects (Hall and Smith, 1955). Van der Valk (1987) pointed out that all changes in wetland plant communities through time are the result of individual species responding to environmental gradients that are largely controlled by water level changes. Specifically, some plant species successfully establish when water levels are lowered and other species are eliminated (Cooke *et al.*, 1986). Van der Valk and Davis (1978) concluded that wetland plant species could be separated into three groups: emergent species such as cat-tail (*Typha glauca*), giant bur-reed (*Sparganium eurycarpum*), and bulrush (*Scirpus fluviatilis*) that germinate on mud flats or in very

118 *Changing resource availability*

shallow water and then persist during flooding; submerged or free-floating species such as duckweeds (*Spirodela polyrhiza*), and water nymph (*Najas flexilis*) that have persistent propagules that survive exposure on mud flats but germinate in standing water; and annual plants that germinate on mud flats but are eliminated by flooding.

(a) Wetlands

Early research on plant succession following drawdown in wetlands initially targeted the performance of plant species that provided food and habitat for ducks. Most of these plant species must have bare sediment in order to become established (Kadlec, 1962). Not only is exposed bare sediment a prerequisite for plant establishment, but the timing and duration of drawdowns are critical. The work of Meeks (1969) in an Ohio wetland demonstrated clearly that as the date of drawdown was delayed each year, so was the gradual development of annual plant dominance (Table 5.7) presumably due to the gradual buildup of an annual plant seed bank. Regardless of the drawdown date, aquatic plant species and semi-aquatic species such as bur-reed (*Sparganium eurycarpum*) and arrowhead (*Sagittaria latifolia*) were gradually replaced by annual weeds such as sow-thistle (*Sonchus* sp.) and touch-me-not (*Impatiens* sp.). As annual weeds gained greater dominance in this wetland, important food and habitat plants for

Table 5.7 Percentage cover of different plant species from 1956 to 1962 in artificially drained Ohio marshes. The drawdown regimen was varied by date but in all cases the marshes were reflooded in September and plant cover was measured in August

Drawdown regimen	Percentage cover						
	1956	1957	1958	1959	1960	1961	1962
Mid-March							
Polygonum lapathifolium	—	16	28	13	12	2	6
Typha sp.	2	3	2	2	2	3	3
Echinochloa walteri	—	16	31	45	3	20	16
Hibiscus palustris	—	1	2	3	13	14	14
Leersia oryzoides	—	1	27	25	41	29	10
Calamagrostis canadensis	—	—	1	4	6	5	6
Semi-aquatic species	—	62	7	1	3	5	6
Annual weeds	0	1	2	7	20	22	39
Nelumbo lutea	36	—	—	—	—	—	—

Table 5.7 Continued

Drawdown regimen	Percentage cover						
	1956	1957	1958	1959	1960	1961	1962
Mid-April							
Polygonum lapathifolium	—	15	21	30	1	1	16
Typha sp.	9	11	12	9	12	14	15
Echinochloa walteri	—	9	10	12	16	6	10
Hibiscus palustris	—	—	—	6	—	1	2
Leersia oryzoides	—	7	30	30	37	21	8
Calamagrostis canadensis	—	1	2	5	4	3	4
Semi-aquatic species	—	57	20	8	10	48	1
Annual weeds	—	—	5	1	20	6	44
Nelumbo lutea	27	—	—	—	—	—	—
Mid-May							
Polygonum lapathifolium	—	44	12	5	—	6	7
Typha sp.	3	14	10	3	2	3	4
Echinochloa walteri	—	2	2	4	—	3	6
Hibiscus palustris	—	—	—	1	1	4	6
Leersia oryzoides	—	1	30	60	76	54	40
Calamagrostis canadensis	—	1	2	3	4	9	11
Semi-aquatic species	—	37	40	17	9	14	3
Annual weeds	—	—	1	7	7	6	22
Nelumbo lutea	10	—	—	—	—	—	—
Mid-June							
Polygonum lapathifolium	—	8	—	6	—	—	—
Typha sp.	8	10	13	13	5	6	6
Echinochloa walteri	—	—	15	28	17	1	6
Hibiscus palustris	—	—	3	20	46	64	43
Leersia oryzoides	—	—	—	6	31	5	38
Calamagrostis canadensis	—	—	—	—	—	—	1
Semi-aquatic species*	—	82	69	26	—	23	5
Annual weeds†	—	—	—	—	—	—	1
Nelumbo lutea	7	—	—	—	—	—	—

* Includes *Sparganium eurycarpum*, *Sagittaria latifolia*, *Scirpus validus*, *Cyperus* sp., and *Eleocharis acicularis*.
† Includes *Sonchus* sp., *Impatiens* sp., *Asclepias incarnata*, *Mimulus* sp., *Bidens* sp., and *Eupatorium perfoliatum*.
From Meeks (1969). Copyright the Wildlife Society. Used by permission.

120 Changing resource availability

Figure 5.5 Various stages of plant community development relative to water-level changes in a Manitoba marsh. In 1980 normal water levels were maintained. In 1981 and 1982 water levels were maintained 1 m above normal. In 1983 the water level was lowered, exposing bare sediments and allowing many annual plant species to colonize. Photograph courtesy of Dr H.R. Murkin.

ducks such as nodding smartweed (*Polygonum lapathifolium*) and Walter's millet (*Echinochloa walteri*) were reduced presumably due to competition from annual weeds. Delaying drawdowns until summer will slow annual plant dominance. Reflooding to depths of about 35 cm eliminates food plant species in about 2 years and cover plants in about 6 years (Harris and Marshall, 1963). Submerged species are most productive and most diverse the first year after reflooding but this drops off with subsequent years of flooding (Figure 5.5). Thus, drawdown is primarily a method of controlling colonization. Reflooding stops colonization, and depending on the tolerances of the individual species to inundation, will selectively eliminate populations through time.

Wetland drawdowns lasting longer than 1 year in duration allow woody plant species to establish and attain considerable size (Harris and Marshall, 1963; Howard-Williams, 1975). This is problematic if the management goal is to provide habitat and food for ducks or if access by boats is to be maintained. Woody plant invasion is of course desirable if the goal is to encourage development of a wetland forest. In general, long-term drainage has a larger and longer-lasting effect on plant community development than does long-term flooding primarily because dense populations of woody plant species become established. For example, Thibodeau and Nickerson (1985) described changes in a Massachusetts wetland after road construction impounded one part of the wetland and drained another part. For six years after road construction significantly more community changes were observed in the drained site than in the flooded site. The drained site showed an increase in the number of species and individuals, primarily woody shrubs and trees such as swamp azalea (*Rhododendron viscosum*), pepperbush (*Clethra alnifolia*), and hazelnut (*Corylus americana*). At the impounded site no new species established and existing species did not expand their populations. Highbush blueberry (*Vaccinium corymbosum*) and winterberry (*Ilex verticillata*), both dominants of the original community, were completely eliminated by flooding.

Both water quantity and water quality can lead to changes in the species makeup of wetland communities. When the water supply is enriched with nutrients, a shift occurs to species with high nutrient requirements. In coastal dunes of The Netherlands nutrient enrichment from river water pumped into the dune systems has caused a reduction in species diversity with the proliferation of a few herbaceous species with high nitrogen requirements such as *Urtica dioica, Eupatorium cannabinum*, and *Epilobium hirsutum* (Van der Meulen, 1982).

(b) Swamp forests

The flooding regimen of a swamp forest is the most important environmental factor controlling community composition and development (Ehrenfeld and Gulick, 1981). Both flooding depth and flooding duration interact in a complex manner to change swamp forest communities through time. When water depths are maintained at about 1.5 m or deeper, most if not all tree species begin to perish (Harms *et al.*, 1980). As the water falls below this depth, individual differences in tolerances of tree species to flooding are expressed. For example, when the Oklawaha River in Florida was impounded, baldcypress (*Taxodium distichum*), tupelo (*Nyssa aquatica*) and cabbage palm (*Sabal palmetto*) were most tolerant to flooding while oaks (*Quercus* spp.), ash (*Fraxinus caroliniana*), and red maple (*Acer rubrum*) were least tolerant (Harms *et al.*, 1980). In general large trees are more flood tolerant than small trees, and mortality due to flooding

Figure 5.6 Changes in a Louisiana wetland forest as a result of modified flooding regimen. (a) The forest community under natural flooding is dominated by baldcypress (*Taxodium distichum*) and tupelo (*Nyssa aquatica*). (b) The forest community that develops when flooded in the autumn and winter and drained in the spring and summer. Baldcypress and tupelo are less important, and red maple (*Acer rubrum*) and ash (*Fraxinus* spp.) dominate the canopy and understorey. (c) This permanently flooded forest has lost most tree species, and buttonbush (*Cephalanthus occidentalis*) has colonized the downed stems. Aquatic plants are also important. Photograph courtesy of Dr W.H. Conner.

tends first to be concentrated in saplings. In short, water-level manipulation controls species performance by sorting species according to inundation tolerance.

Whenever the natural flooding cycle of a swamp forest is artificially modified, changes in the forest community can be expected. Artificial flooding regimes may eliminate the existing tree species in three ways. Existing trees may be killed, reproduction may be stopped, or competing tree species may become established. If the natural rhythm of periodic flooding is replaced by permanent flooding, then retrogressive changes will occur and the community will tend toward less tree dominance and increased dominance of aquatic plants. Downed and dead tree stems are an important structural component of permanently flooded forests. On the other hand, when a regular dry period is imposed, more mesic tree species will establish and eventually usurp typical swamp forest species. Such changes were noted in several Louisiana swamp forests 20 years after the flooding regimen was modified (Figure 5.6). Conner *et al.* (1981) found that the site with an artificial spring/summer dry period, became dominated by red maple (*Acer rubrum*) and ash (*Fraxinus* sp.) while cypress and tupelo decreased in importance (Figure 5.6). In the permanently flooded site, tupelo and baldcypress started dying and buttonbush (*Cephalanthus occidentalis*) expanded into the overstorey by colonizing stumps and downed logs. Clearly, swamp forests may show a certain degree of stability when their natural flooding cycle is maintained. However, if this cycle is artificially modified, a reorganization of the community will ensue. The data again suggest that drainage may have a more profound and longer-lasting effect on community structure than does permanent flooding. Permanent flooding inhibits tree reproduction and accelerates mortality (i.e., controlled species performance); periodic flooding allows new tree species to become established (i.e., controlled colonization).

(c) Shorelines of large impoundments

In large lakes, water depth fluctuations in addition to the disturbing forces of waves must be considered in the development of plant communities at lake margins. Lake edges, like wetlands will typically support monospecific zones of plant species; the processes producing and maintaining these zones are not completely understood (Spence, 1982). Nevertheless, manipulation of the water level is still an important management technique, one that can lead to predictable community changes on shores.

Strongly and moderately exposed shores of reservoirs are subject to intense erosion of the humus layers whereas in sheltered areas no such erosion occurs. This has important consequences for succession at reservoir margins. Nilsson (1981) identified three different types of shore substrates on the margins of a Swedish hydro-electric reservoir that supported different plant communities. Intact humus supported a large number of species and was dominated by the dwarf shrubs *Empetrum hermaphroditum* and *Vaccinium vitis-idaea*. Communities on till that accumulated farther down the shore were less diverse and fine-grained sediments still farther down the shore supported a sparse herbaceous plant community. On till or sediments most of the colonizing plants were annuals with short lifespans and high seed output. Their performance and presence was controlled by the period of drawdown and then they perished when submerged. Plant communities on humus were remarkably stable whereas plant communities on till and sediment were remarkably unstable and changed from one drawdown to the next. Thus at lake margins two types of successional pathways are present. If exposure is low and the humus largely remains intact, plant communities may change slowly over many years; high exposure and the periodic withdrawal of water from till and sediment creates a plant community in a perpetually arrested state of very early succession. Strong fluctuations in the water level (Figure 5.7) will create more areas of early successional communities on exposed till or sediment while stable water levels will eliminate these communities entirely. If water-level drawdowns are regular management activities, then early successional communities on mud flats or shorelines must be considered as permanent landscape elements.

SUMMARY

Changing the duration, level and time of nutrient availability by fertilization can lead to distinct changes in plant community development. The most dramatic responses occur with nitrogen.

Figure 5.7 The shoreline of a Swedish hydro-electric reservoir during (a) summer and (b) winter. Colonization of the exposed sediments is largely by annual species that perish when reflooding occurs. Photographs courtesy of Dr C. Nilsson.

Successional development of old-field communities may be changed by fertilization so that annual-plant dominance is prolonged and perennial-plant dominance is postponed. High rates and durations of nitrogen fertilization in heathland and in plant communities on disturbed soils commonly favour aggressive perennial grasses to the near exclusion of herbaceous species or shrubs. Fertilization of natural grasslands may cause a shift from warm-season species to cool-season species, but this can be avoided by fertilizing late in the growing season. Nutrient exhaustion, the opposite of fertilization, is a largely untested management technique that could be used to favour plant species adapted to low nutrient environments.

Irrigation alone is a short-lived and largely ineffective succession management technique except perhaps in arid grasslands where warm-season grass species are favoured by irrigation. However, when irrigation is combined with fertilization a strong interaction occurs. Some species are favoured and others are inhibited.

Water-level manipulation is a powerful method of controlling succession in wetlands and reservoirs. Drawdown, or the lowering of the water level, can be complete or incomplete. Repeated complete drawdown may lead to the dominance of annual plant species on bare sediments and the elimination of semi-aquatic species. As such, partial drawdowns are recommended to discourage weed dominance and to encourage the development of plant communities of value to wildlife.

In wetlands dominated by woody plant species, draining accelerates succession and increases diversity. Permanent flooding of established forests set in motion a retrogressive succession dominated by selective mortality and the inhibition of reproduction. Tree mortality resulting from flooding varies with water depth, species, and trunk size.

6

Methods of managing succession: changing propagule availability

6.1 INTRODUCTION

The destruction of plant communities in concert with intense soil disturbance such as that caused by mining, construction activities, and off-road vehicle traffic necessitates revegetation or community reconstruction. When these disturbances occur, the propagule banks are often completely removed or destroyed. Revegetation is managed primary succession that relies strongly on augmenting the propagule supply. Successional pathways after revegetation are influenced by physical and chemical conditions of the substrate, methods and timing of revegetation, types of plant species introduced during seeding operations, cultural practices maintained during community development, and the ingress of volunteer species from surrounding plant communities.

In the revegetation of drastically disturbed lands, succession management should operate at many stages of plant community development. Although in practice, much if not all of the effort is concentrated on the early stages such as seedling establishment and the first few years of plant growth, later stages of succession in revegetated communities have been studied in a descriptive sense (Hall, 1957; Gibson *et al.*, 1985; Roberts *et al.*, 1981; Schuster and Hutnik, 1987), but little concerted effort has been devoted to managing the later stages.

This situation is in part the result of environmental regulations that require rapid establishment of persistent plant communities to control erosion following severe soil disturbance; the long-term development of these communities, either progressive or retrogressive is ignored. For example, the Surface Mining Control and Reclamation Act of 1977 in the US has led to widespread use of grasses and legumes in the revegetation of reclaimed surface mine soils in eastern US coal mining regions (Lyle, 1988). Invasion of trees onto these lands is now largely an uncontrolled process, a management option that may or may not be the best method of natural resource use. Resource managers are now attempting to speed the invasion of trees into revegetated landscapes in an effort to develop marketable lumber crops, but they find that the intense competition from grasses and legumes makes tree establishment a difficult process indeed.

This chapter will consider some of the accepted methods of revegetation including site preparation, mulching, seeding, and topsoiling. For a complete treatment of such techniques readers are referred to Schaller and Sutton (1978) and Bradshaw and Chadwick (1980). Most important, however, this chapter will consider succession trends after revegetation and some of the methods that can be used to modify successional trends within the plant community background required by current reclamation laws and regulations. A basic premise of this chapter is that long-term succession can be influenced by propagule input (i.e., controlled colonization) and cultural activities (i.e., controlled species performance) occurring immediately or soon after soil disturbance (Miles, 1988).

6.2 SEEDING AND MULCHING

6.2 Seeding methods

Artificial seeding of disturbed soils can be accomplished by drilling, broadcasting, or hydroseeding. Drilling, where seeds are dropped into a shallow furrow, and broadcasting, where seeds are scattered on the soil surface, are seeding methods limited to relatively flat areas on which tractors can safely manoeuvre. Hydroseeding, a technique where a slurry of seed, binder, fertilizer, and mulch are sprayed from a high-pressure hose, is most commonly used to seed steep slopes. The success of any one method of seeding in a particular region will depend on site properties, physical and chemical characteristics of the soil, environmental conditions after seeding, and the types of plant species used in the seed mixture.

Data indicate that the choice of a seeding method can favour the establishment of different species even though similar seed mixtures are used. In general, forbs tend to establish better when drill-seeded as compared to broadcast seeding. This is especially true when forbs have large seeds (Depuit *et al.*, 1980; Doerr and Redente, 1983). Grasses, on the other hand, establish equally well regardless of seeding method.

A hydroseeding technique was developed by Sheldon and Bradshaw (1977) for the establishment of grasses and legumes on sand slopes. They recommended the slurry contain the following components: 80 kg ha^{-1} seed, legume inoculum, 2.5 t ha^{-1} of *Sphagnum* peat mulch, 3000–5000 kg ha^{-1} of limestone and 125 kg ha^{-1} of 17 : 17 : 17 fertilizer. Sheldon and Bradshaw found that fast-release fertilizers applied in excess of 125 kg ha^{-1} actually inhibited seed germination. They suggested that mulch be applied to offset this effect or that fertilizers should be applied either before or after seed germination occurs.

A high success rate with any method of seeding requires the creation of numerous safe sites for seed germination. Such safe sites can be maximized by roughening the soil surface before seeding. Ridges, clods, cracks, and grooves tend to hold and facilitate germination better than smooth graded surfaces (Wright *et al.*, 1978). The features of a rough soil apparently increases the soil/seed contact surface thus enabling better water uptake.

6.2.2 Mulches

Mulches are typically used in combination with seeding to retard soil erosion, improve the germination of seeds, conserve soil moisture, and ameliorate soil temperatures. Mulches used in revegetation include straw, hay, wood chips, shredded bark, peat moss, corncobs, sewage sludge, and synthetic petroleum products. The value of mulches as deterrents of soil erosion on bare soil is not questioned (Slick and Curtis, 1985); indeed, mulching is required by law in many countries following severe soil disturbance. However, mulches may break the continuity between soil and seed thus leading to seedling desiccation or they may release chemicals that inhibit seed germination (Naylor, 1985). It is difficult to separate beneficial effects of mulch on soil erosion from the negative effects on seed germination.

Both the rate of mulching and the type of mulch may influence what species establish as well as the overall success of plant establishment during revegetation. When soil erosion is a problem on steep slopes, mulched sites have a higher rate of plant establishment than unmulched sites (McGinnies, 1987; Ringe and Graves, 1987). This is presumably the result of mulches holding seeds in place and also conserving soil moisture. Selection for certain species at the establishment phase by using different mulches is not well studied, and the data are contradictory from site to site. For example, in a long-term study of different mulches on Kentucky coal mine spoils Dyer *et al.* (1984) found that during the first growing season after seeding, bark or hay mulching gave about twice as much vegetative cover as controls with no mulch, while other types of mulches did not increase plant cover relative to controls (Table 6.1). By the seventh growing season after seeding, none of the mulched plots had significantly more grass and legume biomass than controls but bark, hay, and wood chip mulching did tend to favour grasses rather than legumes (Table 6.1). At an

Table 6.1 Biomass (kg ha^{-1} oven dry weight) of grasses and legumes growing on a Kentucky coal mine spoil 7 years after seeding and the application of various types of mulches

Treatment	Mulch application rate	Biomass (kg ha^{-1})	
		Grass	Legume
Petroset	378 l ha^{-1} + 560 kg ha^{-1} hydromulch*	605	13910
Chips	95 m^3 ha^{-1}	683	10068
Straw	3360 kg ha^{-1}	448	13820
Hydromulch	1680 kg ha^{-1}	1142	15019
Bark	95 m^3 ha^{-1}	1377	9184
Hay	3360 kg ha^{-1}	1019	9106
Control	—	616	15098

* Hydromulch is a woody fibre product composed largely of cellulose.
From Dyer *et al.* (1984).

Alabama mine spoil, legumes rather than grasses were favoured by bark and hay mulching (Dyer *et al.*, 1984). This study demonstrates quite clearly that the positive effects of mulching on seedling establishment are short lived and with some mulch types non-existent. However, with all mulches it was shown that soil erosion was slowed as compared to plots without mulches. The effect of mulches on plant establishment needs further study to determine how different plant species at the germination stage are being either encouraged or eliminated.

Tacking hay or straw mulches with either chemical tacking agents such as asphalt or wood fibre provides a means for binding and holding mulches on steep slopes. Such agents must be used cautiously because some release chemicals that inhibit seed germination (Sheldon and Bradshaw, 1977). Wright *et al.* (1978) highly recommended wood fibre (840 kg ha^{-1}) as a tack to straw on steep slopes because of its stabilizing properties and lack of toxic breakdown products. However, they suggested that soils that are stair-step graded, artificially roughened, or fairly level may be successfully revegetated by mulching without tacking.

6.2.3 Seed mixtures

The choice of a seed mixture for revegetation purposes is dependent on management goals and on the availability of commercial seed.

Seeding and mulching 131

Throughout Europe and the US where soil stabilization is a primary management goal for disturbed soils, grass/legume mixtures are widely used (Wright *et al.*, 1978). A number of aggressive grass species are now commercially available; among these species, cultivars can be obtained that are adapted to specific environmental conditions (Table 6.2). Seeding grasses with legumes or seeding legumes in monospecific stands (Table 6.3) is an efficient method for rapid soil stabilization and

Table 6.2 Aggressive, cultivated grasses that can be seeded on disturbed soils for rapid stabilization. These species will arrest succession on some sites, especially if provided with artificial nitrogen fertilizers. Recommended seeding rates are approximate and will vary with seed mixture

Climate*	Common name	Scientific name	Seeding rate ($kg\,ha^{-1}$)
T	Weeping lovegrass	*Eragrostis curvula*	5–10
T	Perennial ryegrass	*Lolium perenne*	25–30
T/CT	Red fescue	*Festuca rubra*	25–50
T/CT	Kentucky bluegrass	*Poa pratensis*	5–30
T/CT	Redtop	*Agrostis gigantea*	30–50
T/WH	Ky-31 Tall fescue	*Festuca arundinacea*	40–60
CT	Orchardgrass	*Dactylis glomerata*	5–15
CT	Hairgrass	*Deschampsia caespitosa*	15–20
CT	Chewings fescue	*Festuca rubra* var. *commutata*	15–20
CT	Canada bluegrass	*Poa compressa*	5–15
WH	Grease grass	*Melinis minutiflora*	10–15
WH	Bahiagrass	*Paspalum notatum*	30–40
WH	Bermudagrass	*Cynodon dactylon*	5–15
SA	Crested wheatgrass	*Agropyron desertorum*	5–10
SA	Siberian wheatgrass	*Agropyron sibiricum*	5–10
SA	Intermediate wheatgrass	*Agropyron intermedium*	5–10
SA	Reed canarygrass	*Phalaris arundinacea*	2–5
SA	Smooth brome	*Bromus inermis*	5–10
SA	Russian wildrye	*Elymus junceus*	5–10
SA	Hard fescue	*Festuca ovina* var. *duriuscula*	5–10

* T, temperate; WH, warm humid; CT, cool temperate; SA, semi-arid.
Wright *et al.* (1978).

Table 6.3 Aggressive, cultivated legumes that can be seeded singly or in combination with grasses for long-term soil nitrogen accumulation on disturbed sites. These species may arrest succession especially if provided with artificial phosphorus fertilizers. Seeding rates will vary with seed mixtures

Climate*	Common name	Scientific name	Seeding rate (kg ha^{-1})
T	Flatpea	Lathyrus sylvestris	20–25
T	Birdsfoot trefoil	Lotus corniculatus	20–25
T/CT	Red clover	Trifolium pratense	5–10
CT	Alsike clover	Trifolium hybridum	5–10
T/CT/WH	White clover	Trifolium repens	5–10
WH	Crimson clover	Trifolium incarnatum	25–30
WH	Desmodium	Desmodium uncinatum	5–10
WH	Sericea lespedeza	Lespedeza cuneata	20–25
WH/T	Crownvetch	Coronilla varia	15–25
SA	Alfalfa	Medicago sativa	5–10
SA	Cicer milkvetch	Astragalus cicer	5–10
SA	Yellow sweetclover	Melilotus officinalis	5–7

* T, temperate; CT, cool temperate; WH, warm humid; SA, semi-arid.
Wright *et al.* (1978).

long-term soil nitrogen accumulation. Where legumes are established in pure stands, rates of soil nitrogen accumulation as high as 180 kg ha^{-1} year^{-1} have been measured on previously disturbed soils (Palaniappan *et al.*, 1979).

The use of perennial grasses and legumes in the revegetation of disturbed soils, especially when coupled with either nitrogen or phosphorus fertilization, may lead to productive plant communities that successfully inhibit soil erosion. But these same plant communities may show low diversity and may also resist vegetation change for many years (Figure 6.1). Management goals of high diversity and high productivity may not be attainable because environmental factors such as fertilization that lead to high productivity inevitably lead to the dominance of a few species (Biondini and Redente, 1986). Nevertheless, resource managers are searching for alternatives to the use of aggressive, non-native grasses and legumes for revegetation purposes in an effort to increase diversity and accelerate successional change on disturbed soils.

Figure 6.1 In the foreground a patch of crownvetch (*Coronilla varia*) planted on mine spoil in eastern Kentucky. This community has resisted invasion by other species for 20 years. Plots in the background planted to other species are now heavily invaded by *Rubus* and *Rhus* spp.. Photograph courtesy of the US Forest Service and Dr Gary Wade.

Native plant species in seed mixtures have great potential for developing diverse plant communities on disturbed soils (Figure 6.2). When native species are seeded, fertilization rates must be kept low and aggressive alien species must be meticulously screened from the seed mixture. Depuit *et al.* (1980) seeded a mixture of native grasses, forbs, and shrubs onto Montana coal mine spoils, fertilized with 30 kg ha^{-1} N and 16 kg ha^{-1} P, and then noted the establishment and persistence of diverse communities dominated by native species.

Because native species are less competitive than introduced species, cover crops sown in tandem with natives must not be aggressive. For example, Gilbert and Wathern (1980) sowed a mixture of native grasses (mostly *Festuca* spp.) onto mine spoils in Great Britain. Compared to grass swards seeded with ryegrass (*Lolium perenne*), the native *Festuca* swards supported a greater rooted frequency of unsown forbs and also covered more of the previously denuded soil surface (Table 6.4). In addition, *Festuca* communities – as opposed to stands

134 *Changing propagule availability*

Figure 6.2 Reconstructed and undisturbed plant communities in north-western Colorado. (a) Graded mine spoil 2 years after seeding with non-native early successional species. (b) Graded minespoil 2 years after seeding and planting with native species, primarily native grasses and big sagebrush (*Artemisia tridentata*). (c) Undisturbed big sagebrush community prior to mining. Photographs courtesy of Dr E.F. Redente.

seeded to ryegrass – were more conducive to the establishment of hand-sown forbs (Table 6.5).

On many disturbed sites, regardless of the original seed mix or fertilization programme, grasses tend to dominate the community after a number of years and forbs tend to decrease in importance (Luken, 1987b; Redente *et al.*, 1984). This is especially true if introduced grass species with strong competitive abilities and vegetative spread by tiller production are in the seed mixture. Exceptions to this generality are situations where woody plant species can establish and persist in competition with the grasses. Table 6.6 shows the results of a long-term study on disturbed soils in Colorado that were seeded with a variety of grasses, forbs, and shrubs. During this study, regardless of initial seed mixture, grasses subsequently dominated the communities because of relatively high seeding rates set by the researchers, higher growth rates of grasses, and the fact that soil conditions at the start of the experiment favoured grasses over shrubs

136 *Changing propagule availability*

Table 6.4 The numbers and rooted frequencies of forb species growing on English mine spoil that naturally invaded grass swards seeded with *Lolium* or *Festuca*

	Lolium	*Festuca*	*Festuca*
Age of sward (years)	5	3	3
Number of unsown forb species (per unit area)	30	36	28
Rooted frequency of unsown forbs	10.6	43.2	109.0
Bare ground (%)	75.6	32.9	27.0

From Gilbert and Wathern (1981). By permission of the Canadian Land Reclamation Association.

Table 6.5 Numbers and rooted frequencies of sown and unsown forb species in *Lolium* and *Festuca* grass swards growing on English mine spoils 5 years after seeding. Grass seeding rate was $15\,\mathrm{g\,m^{-2}}$; forb seeding rate was 10 seeds $\mathrm{m^{-2}}$

	Control	*Lolium*	*Festuca*
Sown forb species (per unit area)	13	12	17
Unsown forb species (per unit area)	11	10	7
Rooted frequency of sown forb species	26.5	26.5	142.0
Rooted frequency of all forb species	53.3	71.6	162.8
Bare ground (%)	79.5	81.2	16.8

From Gilbert and Wathern (1981). By permission of the Canadian Land Reclamation Association.

Table 6.6 Percentage composition of plant species in communities growing on disturbed soils in Colorado 3, 4 and 5 years after seeding with different seed mixtures

Seed mixture*	Species	Composition (%)			
		Seed mixture†	1978	1979	1981
A	*Agropyron* spp.	61	73	74	80
	Stipa viridula	23	6	6	10
	Oryzopsis hymenoides	16	16	10	2
B	*Agropyron* spp.	70	73	68	53
	Elymus junceus	30	24	29	44

Table 6.6 *Continued*

Seed mixture*	Species	Composition (%)			
		Seed mixture†	1978	1979	1981
C	*Agropyron* spp.	29	80	84	83
	Stipa viridula	16	4	4	9
	Oryzopsis hymenoides	9	9	5	1
	Hedysarum boreale	2	—	—	—
	Coronilla spp.	5	—	—	—
	Linum lewisii	13	—	—	—
	Penstemon palmeri	28	—	—	—
D	*Agropyron* spp.	36	55	56	42
	Elymus junceus	24	39	37	45
	Medicago sativa	10	1	2	6
	Sauguisorba minor	15	2	1	2
	Astragalus cicer	13	1	—	1
	Saponaria officinalis	3	—	—	—
E	*Oryzopsis hymenoides*	24	19	12	2
	Agropyron spicatum	18	23	23	53
	Agropyron smithii	32	16	15	14
	Coronilla spp.	7	—	—	—
	Hedysarum boreale	2	—	—	—
	Cowania sp.	2	—	—	—
	Ephedra viridis	4	—	—	—
	Atriplex canescens	9	31	36	17
	Ceratoides lanata	4	3	3	3
F	*Stipa viridula*	24	6	4	6
	Agropyron spicatum	19	11	9	14
	Agropyron desertorum	24	46	51	45
	Agropyron trichophorum	12	30	27	26
	Astragalus cicer	10	—	—	—
	Hedysarum boreale	2	—	—	—
	Cowania sp.	4	—	—	—
	Ephedra viridis	2	—	—	—
	Ceratoides lanata	4	2	3	3

* A, native grasses; B, introduced grasses; C, native grasses and forbs; D, introduced grasses and forbs; E, native grasses, forbs and shrubs; F, native grasses, introduced grasses, forbs and shrubs.
† Plots were seeded with this seed mixture in 1976.
From Redente *et al.* (1984).

138 Changing propagule availability

and forbs (Table 6.6). In general, all seeded forbs did not persist, but two shrub species, fourwing saltbush (*Atriplex canescens*) and winterfat (*Ceratoides lanata*), did manage to persist when seeded. This was attributed to their fast growth rates and an inherent lack of seed dormancy (Redente *et al.*, 1984).

On some revegetated soils it is desirable to accelerate succession to a tree- or shrub-dominated stage (Figure 6.3). This can be accomplished by direct seeding, although establishment of trees from seed requires

Figure 6.3 Three-year-old alders (*Alnus*) planted on graded, fertilized mine spoil in eastern Kentucky. Trees were planted on bare soil. Herbaceous species surrounding the trees are natural colonizers. Photograph courtesy of US Forest Service and Dr Gary Wade.

special management activities not necessary in the seeding of herbaceous species. Successful direct seeding of trees often requires control of seed-eating animals, the preparation of a seedbed with bare mineral soil, and if possible, a sparse cover crop of herbaceous plants (Smith, 1986).

The choice between dormant and non-dormant tree seed is critical if seeds are broadcast or drilled. Using dormant seed is recommended so that germination is staggered or cued to the proper environment for seedling establishment. Artificially breaking the dormancy of seeds by scarification or other methods may assure rapid germination, but if the proper conditions for seedling establishment are not present the seeding may fail. In West Germany, dormant seed of 16–40 woody plant species is broadcast or drill-seeded onto roadside embankments (Luke et al., 1982). This method is apparently successful in establishing both early and late successional tree species in a cover matrix of slow-growing legumes.

Table 6.7 Interactions between types of herbaceous cover and fertilization on the density (number 0.002 ha^{-1}) and height (cm) of black locust (*Robinia pseudoacacia*) seedlings established from seed on Kentucky coal mine spoils. Black locust was broadcast seeded (3.4 kg ha^{-1}) during March 1968 and measurements were taken in the autumn of 1968. Fertilization rates were 67 kg ha^{-1} N and 49 kg ha^{-1} P

Introduced herbaceous species	Herbaceous cover (%)		Black locust seedlings			
			Density		Height	
	NF	F	NF	F	NF	F
Control (no seeding)	3	25	51	48	9	43
Lespedeza stipulacea and *L. striata*	15	50	24	25	5	25
Eragrostis curvula and *Festuca arundinacea*	10	65	24	16	6	17
Festuca arundinacea	15	60	20	8	8	14
Lolium multiflorum	15	60	25	5	7	12

NF, not fertilized; F, fertilized.
From Vogel and Berg (1973). Used by permission of Gordon and Breach Science Publishers Inc.

Competing vegetation sown during revegetation can interfere with the establishment of trees from seed. Because many introduced grasses and legumes are fast-growing, they establish a dense canopy and extensive root systems that successfully compete with tree seedlings for light and soil nutrients. The best approach to assure no competitive inhibition of tree seedlings is to avoid planting a cover crop. Black locust (*Robinia pseudoacacia*) seeded on Kentucky mine spoils performed best when no concomitant seeding of grasses or legumes took place (Table 6.7). However, this is often not possible because of reclamation regulations that require complete soil coverage by plant biomass. An acceptable approach is to plant less-aggressive plant species concurrently. For example ryegrass (*Lolium multiflorum*) inhibited establishment and growth of black locust more than lespedezas (*Lespedeza stipulacea* and *L. striata*) (Table 6.7). The lespedezas are warm season species that are most productive later in the growing season after black locust is established.

Another approach that can improve the success of tree establishment on disturbed soils is to vary the tree planting time relative to the planting of the cover crop and also to combine this with herbicide application. Survivorship of trees is generally higher when they are planted concurrently with the cover crop than when planted after the cover crop is established (Ashby *et al.*, 1988). Spraying a localized spot around the seed planting hole with herbicide improves tree survivorship (Table 6.8) as does mulching (Wittwer *et al.*, 1979).

If the succession management goal involves speeding succession to a tree-dominated phase, decisions must be made whether to plant tree seeds or tree seedlings. In general, trees established as seedlings will achieve greater size over a shorter period of time than trees established from seed (Ashby *et al.*, 1988). Successful planting of seedlings involves many of the same site preparations involved in successful seeding operations such as control of competing vegetation and mulching. Most importantly, however, mineral nutrient deficiencies and toxicities must be corrected.

As with direct seeding, removal of competition from herbaceous plants is beneficial to growth and survival of seedlings. Care must be taken so that seedlings are not damaged by techniques used to control the cover crop. Herbicide directed at the ground and sprayed in a 1.5 m circle around tree seedlings is a proven method of improving seedling survivorship (Table 6.8). Drift effects of the herbicide can be avoided by placing plastic covers over the seedlings during spraying (Fung, 1986). If herbicides cannot be used, scraping the soil around tree seedlings is often successful as a method for controlling the cover crop.

Table 6.8 Effects of tree planting time (concurrent with grass and legume seeding vs. delayed planting 1 year later), planting types (seed vs. seedlings), and herbicide application (sprayed around the planting site vs. not sprayed) on the survival (%) and height (cm) of three species of trees planted on mine spoils in Illinois. Measurements were taken either 5 or 6 years after planting

Species*	Planting	Treatment	Survival (%)		Height (cm)	
			Con-current	Delayed	Con-current	Delayed
FP	seed	control	0.0	0.0	0.0	0.0
FP	seed	herbicide	0.4	0.0	4.2	0.0
FP	seedling	control	75.4	36.2	66.2	41.7
FP	seedling	herbicide	68.8	31.2	93.4	62.6
JN	seed	control	8.8	12.5	20.0	19.0
JN	seed	herbicide	21.7	22.1	46.8	39.4
JN	seedling	control	2.1	1.7	8.1	4.2
JN	seedling	herbicide	15.0	7.1	25.5	21.6
QR	seed	control	2.9	2.1	9.7	4.2
QR	seed	herbicide	7.9	2.9	16.4	7.8
QR	seedling	control	16.2	0.8	24.0	1.9
QR	seedling	herbicide	29.6	5.0	38.4	8.2

* FP, *Fraxinus pennsylvanica*; JN, *Juglans nigra*; QR, *Quercus rubra*.
From Ashby *et al.* (1988). Used by permission.

Planting or seeding early successional tree species does not necessarily guarantee that extended forest development will occur. If sites are well-isolated from forest communities there may not be a sufficient supply of propagules to establish populations of late-successional trees. Therefore, it may be necessary to augment the propagule pool of late-successional species at some later point in time.

6.3 TOPSOILING

Following severe soil disturbance, topsoil is often spread on the spoil surface in an effort to improve the rooting environment for seeded or planted species during revegetation. Topsoil used in such a manner can originate from the surface soils of the area that is to be disturbed. During mining or other activities the topsoil is stock-piled and then redistributed. Topsoil used in revegetation can also be fresh-stripped

142 *Changing propagule availability*

from adjacent plant communities. In addition to the obvious positive effects of topsoil on the rooting environment, topsoil is also a source of propagules that may become established and subsequently make a large contribution to succession on revegetated land surfaces (Figure 6.4). Egler's 'initial floristics' interpretation of succession (Chapter 1) would suggest that the trajectory of succession on topsoiled sites is largely a function of the propagules in the topsoil. However, topsoils from forests are typically dominated by herbaceous species and early-successional trees (Wade, 1989). Therefore propagule input from nearby living trees must also occur if long-term forest succession is to be set in motion.

The value of topsoil in improving the rooting environment does not change as a result of stockpiling. However, the viability of propagules through time does change. The longer topsoil is stockpiled, the fewer

Figure 6.4 Effects of applying forest topsoils to mine spoils in eastern Kentucky. (a) Forest topsoil was collected from a hardwood stand and 1 cm was applied over graded and fertilized fill materials. After 1 year the area now dominated by native species contrasts sharply with the artificially seeded non-native grasses and legumes in the foreground and background. (b) Topsoil was collected from a pine–mixed hardwood forest and 2 cm was applied over graded and fertilized fill materials. In the second year the dominance of black locust (*Robinia pseudoacacia*) contrasts with the non-native grasses and legumes seeded in the foreground and background. Photographs courtesy of the US Forest Service and Dr Gary Wade.

viable propagules remain (Hargis and Redente, 1984). This makes fresh-stripped topsoil an extremely valuable commodity. A technique such a double-stripping that involves covering a deep layer of stockpiled topsoil with a thinner layer of fresh-stripped topsoil is an

144 Changing propagule availability

attempt to stretch the propagule-rich topsoil as far as possible (Tracey and Glossop, 1980). Much recent research has focused on the uses of fresh-stripped topsoil in revegetation and restoration of disturbed soils.

Plant communities that develop from topsoiling will vary depending on the number of propagules in the soil and the regional climate. In arid regions, topsoiling without the supplementation of the propagule supply may increase diversity of early successional communities, but plant cover may not be high enough to satisfy reclamation regulations (Howard and Samuel, 1979). In temperate regions, topsoil, especially

Table 6.9 Biomass (B in kg ha^{-1}) and density (D in number ha^{-1}) of plants emerging from eastern Kentucky forest topsoils under three different seeding treatments

Species	Treatment 1*		Treatment 2		Treatment 3	
	B	D	B	D	B	D
Phytolacca americana†	9	1 874	2053	134 933	245	93 703
Erechtites hieracifolia†	—	—	1965	170 540	438	101 199
Eupatorium serotinum†	—	—	270	110 570	46	127 436
Robinia pseudoacacia‡	4	3 748	50	13 118	4	11 244
Liriodendron tulipifera‡	—	—	17	129 310	4	54 348
Lolium multiflorum§	1859	594 078	—	—	460	198 651
Eragrostis curvula§	745	978 261	—	—	1149	910 795
Festuca arundinacea§	628	732 759	—	—	636	655 922
*Lespedeza cuneata***	32	372 939	—	—	398	1 062 594

* Treatment 1, soils were sterilized and seeded with grasses and legumes; treatment 2, soils were unsterilized and not seeded; treatment 3, soils were unsterilized but were seeded with grasses and legumes.
† native herbs; ‡ native trees; § introduced grasses; ** introduced legume.
From Wade (1989).

that taken from forests, can provide propagules from many different plant growth-forms, including woody species. Seedling establishment when forest topsoils are spread over mine spoils is high, and in many cases no supplementation of propagules is needed to achieve complete plant coverage (Farmer *et al.*, 1982). If seeds are added to the topsoil, it is again imperative that plant species be chosen carefully so that introduced species do not outcompete native species. For example, Table 6.9 clearly shows that the introduced species assumed dominance when seeded into communities developing from Kentucky forest topsoils. If topsoiling produces a complete vegetation cover then

Table 6.10 Percentage cover of introduced plant species seeded on different topsoil treatments that were established over mine spoil in Colorado. Data are presented for plant communities that developed 2 and 5 years after seeding

Species	Percentage cover							
	Treatment 1		Treatment 2		Treatment 3		Treatment 4	
	Year 2	Year 5	Year 2	Year 5	Year 2	Year 5	Year 2	Year 5
Grasses								
Agropyron desertorum	13	14	30	32	26	26	36	38
Agropyron elongatum	7	0	6	0	3	0	0	0
Agropyron intermedium	14	9	15	30	34	17	31	33
Agropyron trichophorum	3	6	2	8	8	26	10	14
Total grasses	47	44	65	85	84	98	95	95
Forbs								
Melilotus officinalis	24	0	23	0	11	0	1	0
Medicago sativa	9	51	4	8	1	2	0	0
Other forbs	20	5	8	7	4	0	4	5

Treatments: 1, 30 cm of topsoil; 2, 30 cm of topsoil plus fertilizer; 3, 60 cm of topsoil over an impervious barrier; 4, 60 cm of topsoil with fertilizer over a barrier.
From Biondini *et al.* (1984/1985).

introduced species should be avoided so successional change can occur as rapidly as possible.

Few studies are available where long-term succession on topsoiled sites is compared to sites not receiving topsoil treatment. Biondini *et al.* (1984/1985) conducted a 5-year study of succession on Colorado mine spoils in an effort to determine the long-term effects of topsoil thickness and fertilization. Plots with shallow topsoils and no fertilization became dominated by *Medicago sativa* after 5 years but all other plots became dominated by grasses (Table 6.10). Among the native species there were no consistent effects of topsoil thickness or fertilization on either successional trajectory or rate. Thus, successional patterns among introduced species appear to be more sensitive to topsoiling and fertilization than successional patterns among native species. If topsoils are used, then efforts should be directed at improving the establishment and performance of native species present as seeds. This can be done by avoiding high rates of fertilization and by not seeding non-native grasses.

6.4 UNMANAGED SUCCESSION FOLLOWING SEVERE DISTURBANCE

Although reclamation laws in both Europe and North America now mandate that severely disturbed soils be revegetated, such laws were not always in place. A number of human-disturbed sites now exist around the world where natural revegetation and succession have occurred without the artificial input of propagules. In addition, areas exist where naturally revegetated and artificially revegetated sites exist side by side, thus providing unique opportunities to study long-term effects of various management activities on succession (Figure 6.5).

Bruns (1986) pointed out that unrevegetated sites produced by various types of mining in West Germany may have intrinsic value because they represent perturbation-dependent plant communities of early successional status. Apparently, such communities and the species in these communities are rare and thus can contribute to landscape diversity (Bruns, 1986). Clearly though, more research is needed before perturbation-dependent communities are accepted as desirable by resource managers or legislators. This is true even though it has been demonstrated that some unrevegetated sites created by mining have high species diversity (Thompson *et al.*, 1986). A survey of natural revegetation from a number of sites may demonstrate what resource managers can expect if human-disturbed sites are allowed to recover without fertilization, mulching, or the artificial input of propagules.

Figure 6.5 Plant communities on a surface mine in eastern Kentucky 20 years after mining ceased. A mixed planting of trees and shrubs is visible in the centre of the photograph. Virginia pine (*Pinus virginiana*) invaded naturally at the edges of the planting as did birch (*Betula nigra*) at the lake edge. The vascular flora contains 350 taxa, 272 of which established naturally (Thompson *et al.*, 1984). Photograph courtesy of the US Forest Service and Dr Gary Wade.

Natural invasion of trees into revegetated sites can be affected by the availability of propagules from surrounding plant communities as well as the types of tree species already growing at a site. Trees with long-range dispersal adaptations are more likely to invade as are species at the edges of disturbed sites. For example, Gibson *et al.* (1985) carried out a study of 49 unreclaimed coal mines in Oklahoma that ranged in age from 10 to 70 years. They found that natural revegetation by trees was a complex process controlled primarily by dispersal mechanisms of tree species. Bird- and wind-dispersed tree species were over-represented in the stripmine communities. Plots planted to early successional trees characteristic of a particular region such as black locust (*Robinia pseudoacacia*) in the eastern US will be more conducive to the entry of other tree species than plots planted to later successional species such as pines (Schuster and Hutnik, 1987).

In contrast to the preceding studies in which volunteer trees readily invade disturbed soils, Hedin (1988) noted two types of natural revegetation pathways on 20 stripmine sites in Pennsylvania ranging in age from 12 to 41 years. One successional pathway leading to a woodland community was characterized by invasion of aspens (*Populus* spp.) and red maple (*Acer rubrum*) into sites that were artificially planted with other tree species. Another pathway, however, was not characterized by volunteer tree invasion and instead an open plant community dominated by lichens (*Cladonia*), moss (*Polytrichum*), and broomsedge (*Andropogon virginicus*) developed. Hedin attributed this lack of tree invasion to extremely low soil pH and to chemical interactions between the ground cover and tree seedlings.

In addition to soil disturbance caused by mining, other types of human-generated soil disturbances set in motion a variety of successional pathways. For example, bulldozed tracks at high elevation sites in Scotland show a distinct successional pathway dependent on elevation (Bayfield *et al.*, 1984). At increasingly higher elevations, bryophytes become more important in natural revegetation, whereas vascular plant species become less important. Plant species found in the tracks were also important components of undisturbed plant communities at the edges of the tracks, once again stressing the importance of surrounding vegetation as a source of propagules.

From these studies, it is clear that unmanaged, severely disturbed soils do not remain as biological deserts for ever. Rather, they function as somewhat unique early-successional sites that express the availability of propagules in nearly plant communities. The question of desirability of such sites is still unanswered. These sites when not managed, may, in some instances, become a source of sediments and pollutants for streams and lakes. Further research is required to establish the economic and ecological tradeoffs that are involved when biological diversity (encouraged by not seeding introduced grasses and not fertilizing) is managed at the expense of erosion control.

SUMMARY

Revegetation is a form of managed primary succession. Propagules of grasses, herbs, shrubs, and trees can be introduced to arrest succession with complete vegetation cover, to provide vegetation cover during natural succession, or to speed succession to a tree-dominated stage.

Because the propagule bank can be destroyed or removed by severe human-generated disturbance, various methods are used to augment the propagule pool. These methods include broadcast seeding, drill seeding, hydroseeding, and direct planting of seedlings. Mulching after seeding or direct planting appears to improve the establishment of many different plant species although it may have a selective effect when seed mixtures are used.

Both introduced and native grass and herbaceous plant species are used successfully in revegetation. Introduced grass species are especially aggressive; native species are generally less competitive. Regardless of whether the seeded community is dominated by introduced or native species, fertilization leads to low diversity and high productivity. Some introduced grass and legume species can, when properly managed, inhibit the invasion of trees, shrubs, and other herbaceous plants.

To accelerate the invasion of trees onto revegetated soils, it is necessary to seed trees at the time of grass and legume plantings or it may be necessary to kill grasses and legumes in the immediate vicinity of the planted seed with herbicides. Planting trees from seed or seedlings into established herbaceous vegetation will probably fail if nutrient deficiencies are not eliminated and competition from herbaceous plants is not controlled.

Spreading of fresh-stripped topsoil onto disturbed sites is a successful method of introducing propagules of native species. If topsoils are stripped from forests and spread on disturbed sites, early-successional trees can become established. Storage of topsoil for long periods of time decreases its value as a source of propagules.

When no attempt is made to revegetate severely disturbed soils, succession will still occur. The types of plant communities that eventually dominate such sites are controlled by chemical and physical characteristics of the disturbed soil as well as by the availability of propagules from surrounding undisturbed plant communities.

7
Animals and succession

7.1 INTRODUCTION

Although ecological succession is often couched in terms of plants, or interactions between plants and the physical environment, numerous facets of succession management include animals. This chapter will consider the role of animals in succession management from two different perspectives: the wildlife biologists manipulating habitats for high and persistent populations of specific animal species, and the land managers who control succession by using grazing animals. One perspective views animal populations changing as a result of succession; the other views animal populations as changers of succession. Both perspectives rely on a thorough knowledge of the animal's life histories and of the interactions that occur between plants and animals through time.

7.2 ANIMAL COMMUNITIES: CHANGES DURING SUCCESSION

Bailey (1984) attempted to group animal species into three successional classes: climax-adapted species, species adapted to developmental stages of succession, and species requiring a mix of successional stages (Figure 7.1). Bailey's classification scheme is useful because it dispels the myth that all animal species are favoured by early successional habitats. It also forces an individual species approach when managing succession.

When managing habitat, and thus succession, for animals, a phase of succession is not the ultimate goal, but rather the biotic and abiotic factors in a successional phase that satisfy resource requirements of target animal species. Consider the thousands of hectares of early successional roadside land created in the US. This land in some parts of the midwestern US could be used by a valuable game bird species such as the ring-necked pheasant (*Phasianus colchicus*). But, due to a combination of human-generated disturbance and a lack of appropriate plant species, it is not suitable pheasant habitat until properly managed (Joselyn et al., 1968).

152 Animals and succession

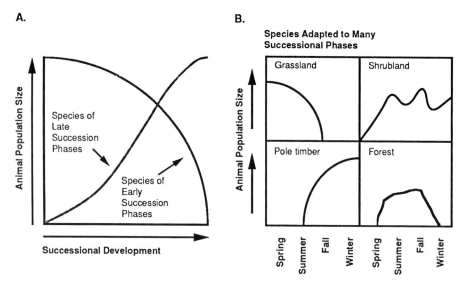

Figure 7.1 (a) Population trends of animal species that require late successional phases or early successional phases. (b) Hypothetical population trends of animal species that require a mixture of successional stages. Such species may utilize different successional phases during different seasons of the year.

Environmental factors considered in succession management for animals include food, water, cover, predation, disease, intra- and interspecific interference, pollution, landscape fragmentation, and others. These factors change during succession as does the suitability of habitat for various animal species (Beckwith, 1954). Superimposed on this relationship between site age and animal communities is the complicating factor of animal mobility. Animal mobility in turn raises critical questions about minimum habitat size, connectedness of habitats, and seasonal habitat use (Forman and Godron, 1986). All in all, management of succession to encourage different animal species is perhaps one of the most challenging tasks facing resource managers.

The first half of this chapter will concentrate on vertebrate species, their population changes during succession, and techniques used to manipulate succession for their benefit. Invertebrate species, although often important as modifiers of succession, are not considered because few situations exist where succession is controlled solely for the benefit of invertebrates.

7.2.1 Birds and old-field succession

During the three decades between 1930 and 1960, old-field succession in the US was studied intensively to determine if bird species also showed patterns of habitat use that coincided with plant succession. In general, succession sorts bird species relative to feeding habits. Early successional phases support granivorous species. In later successional phases these are replaced by insectivorous, and eventually forest and shrubland support mainly fructivorous species (Beckwith, 1954). However, food availability alone does not adequately define suitable habitat.

Some bird species in old-fields have narrow habitat requirements and are found almost exclusively in a single successional phase; still others have broad habitat requirements and are found in several phases. For example, Table 7.1 shows that the yellowthroat (*Geothlypis trichas*) is found exclusively in the grass–shrub phase whereas the cardinal (*Richmondena cardinalis*) is found in three phases. These habitat preferences probably reflect feeding as well as nesting or behavioural requirements of the different species.

From early successional to deciduous forest phases of old-field succession, both bird density and diversity increase (Figure 7.2). The exception to this trend occurs when succession includes a pine-dominated phase. In the US (Johnston and Odum, 1956) and Great Britain (Lack and Venables, 1939) pine forest, due to a lack of structural complexity, tends to support fewer bird species.

7.2.2 Birds and secondary forest succession

There is perhaps more known about the changes in bird communities during secondary forest succession than in any other type of plant community. This abundance of information is due to resource management goals of many state and federal agencies that attempt to maintain both economically viable forest rotation schedules and wildlife populations for recreational purposes. Often, the goal of wood production conflicts with wildlife management goals. In a surprising number of cases, however, secondary succession initiated by forest harvest does benefit certain bird species (Holbrook, 1974).

An overwhelming number of studies suggests that both species diversity and population sizes of birds are increased in regenerating forests created by logging when compared to virgin or uncut stands (Hagar, 1960; Webb *et al.*, 1977). The greater the intensity of tree removal, the greater is the bird response (Webb *et al.*, 1977). Not all

154 Animals and succession

Table 7.1 Breeding bird density in various phases of succession. Plant communities developed as old-fields in the Georgia piedmont. Numbers are the estimated pairs per 40 ha

Succession phase Age (years) Species	Grass–forb 1–3	Grass–shrub 15–20	Pine forest 25–100	Oak–hickory forest 150+
Grasshopper sparrow	40	25	—	—
Eastern meadowlark	15	17	—	—
Field sparrow	—	83	36	—
Yellowthroat	—	33	—	—
Yellow-breasted chat	—	21	—	—
Cardinal	—	9	53	23
Eastern towhee	—	13	53	—
Bachman's sparrow	—	8	10	—
Prairie warbler	—	6	6	—
White-eyed vireo	—	8	9	—
Pine warbler	—	—	148	—
Summer tanager	—	—	47	10
Carolina wren	—	—	29	10
Carolina chickadee	—	—	12	5
Blue-gray gnatcatcher	—	—	15	13
Brown-headed nuthatch	—	—	7	—
Blue jay	—	—	13	5
Eastern wood peewee	—	—	11	3
Ruby-throated hummingbird	—	—	19	10
Tufted titmouse	—	—	16	15
Yellow-throated vireo	—	—	8	7
Hooded warbler	—	—	33	11
Red-eyed vireo	—	—	13	43
Hairy woodpecker	—	—	4	5
Downy woodpecker	—	—	3	5
Crested flycatcher	—	—	11	6
Wood thrush	—	—	6	23
Yellow-billed cuckoo	—	—	1	9
Black and white warbler	—	—	—	8
Kentucky warbler	—	—	—	5
Acadian flycatcher	—	—	—	5

From Johnston and Odum (1956). Used by permission of the Ecological Society of America.

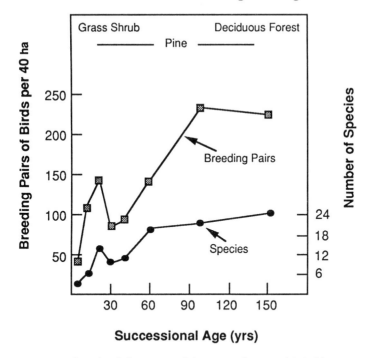

Figure 7.2 Trends in bird density and diversity during old-field succession in Georgia. The dip in both lines is the result of pine invasion. From Johnston and Odum (1956). Used by permission of the Ecological Society of America.

bird species, though, are favoured by logging. The bird response to logging is individualistic and can be positive, negative or neutral (Table 7.2). Positive responses to logging are the result of greater habitat complexity in regenerating forests, especially the development of tall shrub and herbaceous layers. Also this vegetation has greater availability of both seeds and insects. Negative responses are typically found among bird species that require cavities in old trees for nesting or a lack of human disturbance. Indeed, each stage of forest succession following logging is associated with a unique assemblage of bird species (Figure 7.3). Forest harvest methods maintaining many stages of succession will maximize bird diversity across the landscape (Thompson and Capen, 1988).

Structural complexity of forests rather than successional status is often a good predictor of habitat quality for birds (Niemi and Hanowski, 1984). Retaining dead trees and shrubs during forest

Table 7.2 Responses of bird species to logging in the Adirondack Mountains of New York

Not affected	Increased	Decreased
Red-eyed vireo	American redstart	Blackburnian warbler
Black-throated blue warbler	Chestnut-sided warbler	White-breasted nuthatch
Swainsons's thrush	Veery	Wood thrush
Scarlet tanager	Broad-winged hawk	Ovenbird
Chimney swift	Rose-breasted grosbeak	Least flycatcher
Eastern wood peewee	White-throated sparrow	Black-throated green warbler
Dark-eyed junco	Canada warbler	Winter wren
Blue jay	Black-and-white warbler	Yellow-bellied sapsucker
Hermit thrush	Black-capped chickadee	

From Webb *et al.* (1977). Copyright the Wildlife Society.

harvest is a simple method of assuring that complexity and a large number of bird species will be maintained during forest rotations.

Overall, there is abundant evidence that secondary successional communities created by logging or tree stand improvement activities do increase both bird populations and bird species diversity as habitat complexity increases. However, some of the more specialized species, ones that require large tracts of undisturbed forest, will gradually be eliminated from an area as logging operations transform landscapes into patches of regenerating forest. Methods for maintaining multispecies assemblages of animal species within a background of various types of forest use will be discussed in Chapter 9.

7.2.3 Birds and the edge effect

Some bird species tend to be more abundant at the edges of forests than in forest interiors (Johnston, 1947; Gates and Gysel, 1978; Strelke and Dickson, 1980). These edge species were formally defined by Johnston and Odum (1956) as species that are common throughout an area of different successional phases, but do not occur solely within a single successional phase. Because patches of habitat are managed for bird

Animal communities: changes during succession 157

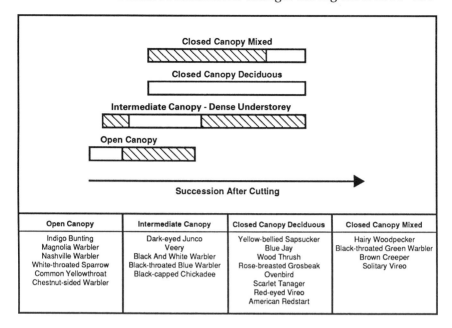

Figure 7.3 Species assemblages of birds found in various successional phases of Vermont hardwood forest following logging. Open bars indicate the presence of an entire assemblage. Diagonally hatched bars indicate the presence of a partial assemblage. From Thompson and Capen (1988). Copyright the Wildlife Society.

species, the unavoidable result is edge. The creation of two adjacent plant communities, each in different successional phases, may actually create three unique habitats (Figure 7.4). Increasing fragmentation of the landscape into patches creates more edge habitat and in turn favours bird species that utilize edges.

Even though the edge effect is well established for birds, and management often strives to maximize edges (Strelke and Dickson, 1980), few studies have attempted to determine what resource factors are supplied by edges. Kroodsma (1984) suggested that the edge effect may result from singing-post preferences of brushland bird species rather than an overall habitat preference. He hypothesized that brushland bird species begin to use forest edges when adjacent brushy vegetation is lacking or when large open areas abut forest. Instead of maximizing edge habitat, resource managers might consider maximizing large blocks of brushy habitat to assure that brush-land species are not restricted solely to forest edges (Kroodsma, 1984).

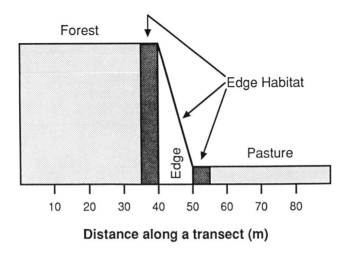

Figure 7.4 Edge habitat created at the junction of a forest patch and a pasture. Edge effects develop in both the forest and the pasture.

7.2.4 Some bird species requiring several successional phases

(a) Ring-necked pheasant

The ring-necked pheasant (*Phasianus colchicus*), a popular introduced game bird in the US is a classic example of a species that requires several successional phases. Pheasants utilize seeds of annual plants from early successional phases primarily in autumn and winter. They seek the cover of perennial grass and perennial herb phases especially in winter, and also use these habitats for nesting in the spring. Shrub-dominated phases serve as roosting areas in all seasons except summer (Beckwith, 1954).

Edminster (1954) suggested that land managed for ring-necked pheasants should include a mixture of cropland, grassland, and woodland (Table 7.3). He suggested other land management practices conducive to pheasant populations: planting supplementary food patches with annual crops, delaying ploughing of corn stubble until spring, delaying mowing of pasture or hayfields in the spring until nesting is complete, and planting cover crops in drainage ditches and on roadsides.

Unfortunately, management recommendations established in the 1950s for game birds did not foresee the dramatic increases in land prices and concomitant changes in agricultural methods that occurred in the midwestern US during the 1970s and 1980s. Associated with monocultures was a corresponding decrease in grassland, woody

Table 7.3 Ideal habitat mix for encouraging populations of ring-necked pheasants. All habitat types should be well-interspersed as patches less than 4 ha

Habitat type	Percentage of land use
Cropland	
Corn	25
Small grains	25
Total cropland	50
Shrub, hedgerows or marsh	5–20
Grass–legume pasture	30–45

From Edminster (1954).

cover, and edge habitats (Vance, 1976). Pheasant populations declined precipitously. However, where special programmes are enacted to limit disturbances by people – for example, protecting roadside grasslands from mowing – pheasant populations do respond positively. This is especially true when protected roadsides are adjacent to small-grain fields (Warner and Joselyn, 1986).

(b) Red grouse

Red grouse (*Lagopus lagopus*), a popular game bird of Europe, are tightly linked to the successional phase of their primary food source, heather (*Calluna vulgaris*). Experimental burning of heather so that burned sites are scattered throughout a matrix of unburned heather increases red grouse populations (Miller *et al.*, 1970), but this population increase is delayed until 3 years after burning and then persists for only 3 years (Figure 7.5). The reasons for this response are not entirely understood, but red grouse do prefer heather that is 2–3 years in age. Older or younger shoots are rejected (Moss *et al.*, 1972). Thus red grouse may be responding to nutritional factors in the heather that change due to burning or as a result of ageing of heather populations. Fertilization of heather also increases grouse populations, supporting the role of nutrients in tissue as a limiting factor for red grouse (Miller *et al.*, 1970).

Cover and nesting sites are provided by heather more than 10 years of age. In addition, openings or thinnings are necessary if grouse are to move freely through the heather. Thus, heather managed for grouse must contain a range of developmental phases. Determining the size,

160 *Animals and succession*

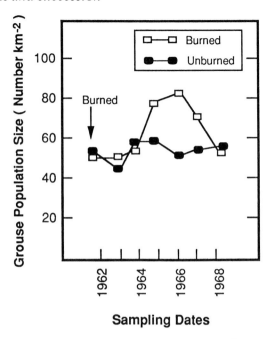

Figure 7.5 Population trends of red grouse (*Lagopus lagopus*) in response to prescribed heather burning. Populations did not begin to increase until three years after burning. From Miller *et al.* (1970).

frequency, and distribution of prescribed burns to maintain red grouse populations is presently a major problem facing heathland managers.

7.2.5 Large vertebrates requiring a range of successional phases

(a) White-tailed deer

White-tailed deer (*Odocoileus virginianus*) are widespread throughout North, Central and South America. They are browsers, feeding largely on woody plants. Habitat manipulations that increase the availability or palatability of woody species are common in white-tailed deer management.

White-tailed deer respond positively to logging if forest regeneration is allowed to occur. A number of studies suggest that browse availability to white-tailed deer is at a maximum 2–3 years following forest harvest (Blair and Burnett, 1977; Dills, 1970). This results from increased release of sprouts and from a more open canopy that allows release of both shrubs and groundlayer species. However,

Table 7.4 Mean amounts (kg ha^{-1} oven dry mass) of available browse in Michigan mixed-conifer forests 8 and 14 years after cutting by various harvest methods. All cuts were made in 1952

Cutting method	Mean amount of browse (kg ha^{-1} oven dry mass)	
	1960	1966
Uncut	10.7	10.7
Selection*	24.4	29.9[a]
Shelterwood†	97.7[a]	32.3[a]
Diameter-limit‡	81.4	38.4
Block§	71.0	44.9
Strip**	92.2[a]	63.5

* Commercially desirable trees removed.
† Trees of lower crown classes removed.
‡ Only trees above a critical diameter were removed.
§ Blocks 0.16 ha in area were clear-cut.
** Strips 23 × 123 m were clear-cut.
Means within a column with the same letters are not significantly ($P < 0.05$) different.
From Krefting and Phillips (1970). Used by permission from the Society of American Foresters.

food and cover requirements need to be satisfied by forest harvest methods.

A variety of forest harvest techniques have been tested in white-tailed deer management. The best approach is to alternate clear-cut strips with uncut strips or patches. Table 7.4 shows that clear-cut strips in Michigan mixed-conifer forest had higher browse availability for longer periods of time than sites subjected to other forest cutting techniques. In addition, deer made heavy use of uncut strips adjacent to the cut strips (Table 7.5). The success of this management practice is the result of close contact between sites that provide food and sites that provide cover.

(b) Moose

The moose (*Alces alces*) is an animal species that requires frequently disturbed boreal forest communities. There are numerous cases in North America (Peek *et al.*, 1976) and Fennoscandia (Markgren, G.

Table 7.5 Numbers of deer pellets ha^{-1} counted in Michigan mixed-conifer forests 8 and 14 years after various types of forest cutting

Cutting method†	Deer pellets (no. ha^{-1})	
	1960	1966
Uncut	74	35
Selection	148	64
Shelterwood	232*	133*
Diameter-limit	158	64
Block	44	20
Strip (cut and uncut)	94	40
Strip (uncut only)	252*	198*

* Significantly ($P < 0.05$) different from the uncut treatment.
† Cutting methods as in Table 7.3.
From Krefting and Phillips (1970). Used by permission from the Society of American Foresters.

1974) where moose populations have increased following logging or fires. However, when logging occurs to the extent that habitat diversity is decreased, the populations decline (Telfer, 1974).

Lands managed for moose should have the following mix of habitats and successional phases: logged forest less than 20 years old should make up 40–50% of the land; mature spruce–fir forests should comprise 5–15% of the land; and the remainder of the land (35–55%) should be distributed among older aspen–birch woodland and aquatic habitats (Peek et al., 1976). Logged areas and aquatic habitats have high browse and forage availability in early summer and late autumn. Spruce–fir forests are used in winter as cover whereas fir may be browsed heavily in winter as a secondary food source. Older aspen–birch stands are frequented in the summer.

7.2.6 Animals requiring a single successional phase

Animals requiring large blocks of very old, undisturbed forest communities pose unique problems for succession managers. Most old-growth forests on earth are now under the threat of various types of person-generated disturbances. It is not surprising that many animal species that require old forests are endangered or threatened.

(a) Spotted owl

The spotted owl (*Strix occidentalis*) is a bird of the western US that requires large blocks of mature (200–500 year old) forest. It is estimated that 400–880 ha of old-growth forest are necessary for a single pair of spotted owls (Carey, 1985). Spotted owl populations decline when the mature forest becomes fragmented by logging activity (Carey, 1985).

Succession management for spotted owls is presently taking a number of paths. As Carrier (1985) pointed out, resource managers are skilled at setting back succession, but few methods are available to speed up succession so that relatively young plant communities resemble old growth. The US Forest Service is mapping potential spotted owl territories in national forests and these will be managed specifically for spotted owls in the future (Carrier, 1985). Spotted owl management primarily involves protection from human-generated disturbances.

(b) Marten

Martens (*Martes americana*) are common inhabitants of mature coniferous forests in parts of North America. Changes in marten populations relative to forest disturbance are well-studied; it appears that martens can co-exist with moderate amounts of forest harvest but they do not thrive near such activity. Soutiere (1979) studied marten populations in Maine on undisturbed, partially cut and clear-cut sites. He found that martens did not use clear-cuts up to 15 years of age. Partially cut forests (basal area reduction of 40%) and undisturbed sites supported similar marten populations, but marten populations on clear-cuts were reduced by 66%. Because martens avoid large openings, Soutiere (1979) suggested that clear-cut patches should be small and well-interspersed with mature or partially cut forest.

7.3 ECOLOGICAL SUCCESSION: CHANGES DUE TO ANIMAL ACTIVITIES

There are several reasons why resource managers are concerned about the effects of animals on succession. First, populations of grazing or browsing animals, either domestic or wild, are common components of most ecosystems. Their presence is mandated by resource management goals that simultaneously provide many land uses (e.g., hunting, livestock production, and nature interpretation). Second, animals can achieve or work against succession management goals. The introduction of grazers can be the simplest solution to a succession 'problem'; too many grazers (i.e., overgrazing) may lead to plant community deterioration and a loss of ecosystem integrity. Last, animal populations erupt and crash, which may in turn explain short-term changes in plant communities. Resource managers need a thorough knowledge of how animal activities affect succession and how the varying of this animal activity changes succession.

Ellison (1960) summarized much research regarding the effects of grazing animals on succession in rangelands. He identified a number of organism, community, and ecosystem characteristics that may be changed by grazing animals, all of which, when modified, will also change successional pathways. Included here are plant growth rates, plant biomass allocation, mulch distribution and abundance, seed production and availability, soil disturbance, erosion and soil fertility (Ellison, 1960). Considering the primary activities of grazing or browsing animals such as plant biomass removal, trampling, defaecating, and others (Figure 7.6), it is not surprising that few generalizations can be made about the effects of animals on succession. What follows, however, is an attempt to summarize data from a variety of plant communities influenced by several common grazing or browsing animal species.

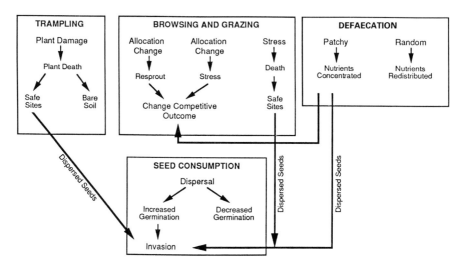

Figure 7.6 Various activities associated with grazing or browsing animals and their possible effects on plants.

7.3.1 Evidence used to assess the effects of animals on succession

Large grazing or browsing animals can be eliminated from specific areas by a number of methods. Most commonly, fencing is used to create an exclosure (Figure 7.7). Animals excluded from one area are by default included in an adjacent enclosure. When succession within exclosures is compared to successional pathways outside the exclosures, one can assume that the differences, if any, are the result of animal activities. Succession within exclosures may be artificial because animals are a common component of most communities (Beetle, 1974). However, the use of exclosures is widespread and it provides a necessary control to understand animal influences.

The role of animals in plant community development can also be determined if for some reason a dominant animal species is eliminated in a specific area. For example, the outbreak of myxomatosis eliminated British rabbit populations in 1954. Ongoing studies of community development in grasslands before, during, and after the outbreak provided a clear picture of the role of rabbits in this system

Figure 7.7 A typical exclosure used to study the effects of rabbit grazing on plant community development in Australia. Note the taller vegetation inside the exclosure. Photograph courtesy of Dr J.H. Leigh.

166 *Animals and succession*

(Watt, 1962). Likewise, removal of large herbivores by poaching in African national parks has caused measurable changes in succession (Smart *et al.*, 1985).

7.3.2 Rabbits

Watt (1957, 1960b, 1962) designed and carried out an incisive series of experiments to determine the effects of rabbits (*Lepus cuniculus*) on grassland succession in England. His results demonstrated clearly that physical environment factors and rabbit grazing can interact in a complex manner to change community development pathways. In two of the grassland communities with relatively deep soils (one acidic, the other slightly alkaline and sandy) protection from grazing allowed the grass *Festuca ovina* to expand in importance and eventually dominate (Watt, 1957, 1960a). However, in grassland A, a community occurring on shallow calcareous soils, protection from rabbit grazing caused an

Figure 7.8 Fenceline separating experimental plots in the Kosciusko National Park, New South Wales. The plot on the left was not burned or grazed by rabbits; the plot on the right was not burned but was stocked with rabbits ($10\,ha^{-1}$). Flowering stalks and flowers of *Hypochoeris radicata* and *Craspedia* spp. are nearly absent from the grazed plots. Photograph courtesy of Dr J.H. Leigh.

increase in the number of species, presumably because rabbits were selectively eliminating these species before exclosure.

When rabbit populations are high, retrogressive changes in plant communities occur. Two kinds of rabbit activities cause this change: foraging and digging. For example, Leigh et al. (1987) showed that subalpine plant communities in Australia with high (10–20 rabbits ha^{-1}) populations had diminished cover of palatable forbs and also more disturbed soil than control areas (Figure 7.8). As long as rabbit pressure is maintained it is likely that the system will assume a new successional trajectory. Physical environment factors such as fire can actually increase rabbit pressure as sprouts of shrubs and forbs are heavily grazed and rabbit populations expand (Leigh et al., 1987).

In arid regions rabbits do not exert strong control over succession perhaps because plant species in the environment are more resistant to biomass removal. Over time periods of 6–15 years Rice and Westoby (1978) did not find measurable differences when Great Basin plant communities inside exclosures were compared to control sites.

7.3.3 Sheep

A large number of studies are now available where ecological succession is measured through time inside and outside sheep exclosures. These studies demonstrate clearly that grazing by sheep can change plant succession and most importantly that plant species respond individualistically to grazing pressures.

Marrs et al. (1988) summarized data from a *Juncus squarrosus* grassland in Great Britain where an exclosure was built to stop sheep grazing. Five different types of plant responses (Figure 7.9) to sheep grazing were described:

1. A number of species such as *Potentilla erecta* and *Nardus stricta* were more abundant in the grazed plot, but among these species some maintained constant importance while others gradually decreased.
2. A few species decreased in importance after protection from grazing (e.g. *Juncus squarrosus*). In some cases declines were measured simultaneously in both plots but usually the rate of decline was fastest in the grazed plot.
3. Some species decreased in importance when grazed (e.g. *Carex nigra*).
4. Some species increased in importance after protection from grazing (e.g. *Calluna vulgaris*).

168 *Animals and succession*

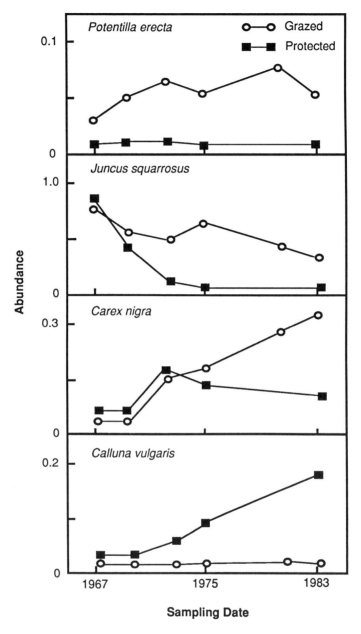

Figure 7.9 Long-term trends in the abundance of different plant species growing in *Juncus squarrosus* grassland in Great Britain when grazed by sheep (1.4 sheep ha^{-1}) and when protected from grazing. Plots were protected in 1966. From Marrs *et al.* (1988). Reprinted by permission of Kluwer Academic Publishers.

5. Some species had curvilinear or inconsistent changes in importance during the course of the study (e.g. *Eriophorum vaginatum*), but usually the direction of change was similar in grazed and ungrazed plots.

Marrs *et al.* (1988) concluded that the protected and unprotected plant communities showed similar successional changes because both showed increased dominance of *Eriophorum vaginatum* and *Calluna vulgaris*. However, the protected community changed faster. Thus sheep grazing may be a viable method of slowing successional change in grassland.

In subalpine communities where tussock and intertussock microsites are common, effects of sheep grazing are concentrated in the intertussock areas because of sheep trampling and higher grazing pressure. Shifts in palatable and grazing-resistant species are common. For example, in Australian subalpine communities palatable forb species such as *Asperula* spp. decreased in grazed plots while hard-leaved shrubs such as *Bossiaea foliosa* and grazing-resistant grasses such as *Festuca asperula* increased (Figure 7.10).

The rapid changes that occur in plant communities under sheep grazing or when sheep grazing is stopped prompted researchers to develop grazing systems for community conversion. Although both grazing pressure and timing of grazing have been manipulated, this has not been an area of great discovery. Sheep grazing systems for converting arable land into species rich grassland (Gibson *et al.*, 1987) and for expanding heathland (Bakker *et al.*, 1983) have been tested. In general, sheep grazing does allow some desirable species to establish and expand by decreasing the dominance of a few aggressive species, and by creating safe sites. However, large accumulations of dung can cause the invasion and spread of aggressive grass species and sheep tend to graze the desirable species as soon as they establish (Bakker *et al.*, 1983; Gibson *et al.*, 1987). The best approach to increasing species diversity in grasslands and heath may be periodic grazing followed by rest.

Sheep browsing has also been tested as a potential method of cropping back shrubs and tree sprouts to encourage plant communities free of woody plants. For example, Lindsay and Bratton (1979) found that sheep did consume sprouts of oak, hawthorn and blueberry on grassy balds in Great Smoky Mountains National Park but also tended to graze desirable grass species heavily. Thus to use sheep successfully for this purpose it may be necessary to limit grazing pressure to the times of the year when grasses are not available or are less palatable.

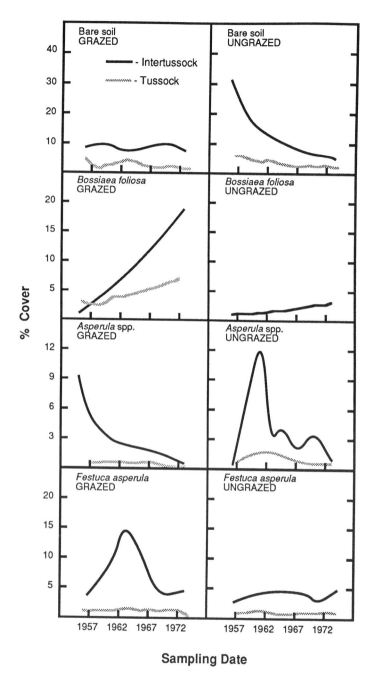

Figure 7.10 Percentage cover trends of various plant species in Australian subapline plant communities when grazed by cattle and sheep (five animals ha^{-1}) or when protected from grazing. Solid lines indicate intertussock microsites; dotted lines indicate tussock microsites. From Winbush and Costin (1979). Used by permission.

7.3.4 Cattle

The effects of cattle grazing on succession are well studied. Indeed, the entire discipline of range management is concerned mostly with grazing management and the responses of plants to various degrees of grazing pressure. However, a survey of the literature reveals a surprising diversity regarding successional trends when cattle grazing is started, when it is stopped, or when it is varied in intensity, and few generalizations can be made. In order to predict how a plant community will respond to grazing by cattle, succession managers must carry out preliminary trials or rely on published data that hopefully have been collected in the plant community type that is being managed.

Moderate cattle grazing tends to increase both species and plant community diversity. In Dutch salt marsh communities this is because grazing creates safe sites for plant establishment and also keeps

Figure 7.11 Experimental cattle exclosure established in 1946 in the Bogong High Plains of southeastern Australia. Species with greater abundance in the exclosure include the silver snow daisy (*Celmisia* sp.) and the dwarf alpine starbush (*Asterolasia trymalioides*) both of which are heavily grazed by cattle. Unpalatable shrubs such as *Hovea longifolia* and *Kunzea muelleri* are expanding throughout the grazed area. Photograph courtesy of Dr R.J. Williams.

aggressive species from expanding in coverage (Bakker and Ruyter, 1981). In the absence of grazing in salt marshes, *Elytrigia* spp. form large clones and dominate the landscape (Bakker and Ruyter, 1981). Heather moorlands of Scotland show increased diversity of graminoids and forbs and decreased importance of *Calluna vulgaris* when heavily grazed (Welch, 1984). Part of this response is due to the transport of seeds in dung and also to the greater abilities of grasses and forbs to emerge from heavy dung accumulations (Welch, 1985).

A number of studies suggest that shrub invasion of heathland and grassland (Figure 7.11) is accelerated by cattle grazing (Williams and Ashton, 1987; Vinther, 1983). Because cattle may also browse on shrubs, it is not uncommon to find that the fastest rate of shrub invasion is when heavy grazing (shrub establishment) is followed by an absence of grazing (shrub growth).

The development of different cattle-grazing systems allows range managers systematically to vary the cattle grazing pressure. Because grazing pressure is one of the most important determinants of plant population response to grazing (Ellison, 1960) succession managers

Table 7.6 Mean percentage cover of shrub, midgrass and shortgrass communities in Texas rangelands after 4–20 years of different cattle grazing systems

Grazing system	Plant communities		
	Shrub	Midgrass	Shortgrass
Exclosure 1	16.8	83.2	—
Exclosure 2	8.4	91.6	—
Deferred rotation (DR*)	10.2	60.4	29.4
Continuously grazed, moderately stocked (CGM†)	8.9	60.2	30.9
High intensity, low frequency (HILF‡)	7.6	48.0	44.4
Continuously grazed, heavily stocked (CGH§)	9.8	23.2	67.0

* Grazed 12 months followed by 4 months of rest, 6.2 ha au^{-1}.
† Continuous grazing, 6.5 ha au^{-1}.
‡ Seventeen days of grazing (6.5 ha au^{-1}) followed by 119 days of rest.
§ Continuous grazing, 4.6 ha au^{-1} (where au = animal unit).
From Wood and Blackburn (1984). *Journal of Range Management* (1984). Used by permission.

should be aware of the different grazing systems and their impact on succession.

Wood and Blackburn (1984) compared plant communities from native rangelands in Texas that were assigned to various types of cattle grazing systems (Table 7.6). The most striking difference among the grazing treatments noted in this study was the percentage of shrub, midgrass and shortgrass communities (Table 7.6). Grazing pressure was clearly highest in CGH and HILF systems because these showed

Table 7.7 Percentage cover of plant species in three communities (shrub, midgrass and shortgrass) of a Texas rangeland after 4–20 years of different cattle grazing systems

Species	Percentage cover					
	CGH*	CGM	DR	HILF	Ex. 1	Ex. 2
Shrub						
Bare ground	7.0	17.6	4.0	7.1	1.0	1.2
Bouteloua curtipendula	3.2	4.5	4.2	2.0	9.0	2.4
Stipa leucotricha	70.1	29.4	71.8	48.3	3.5	3.2
Buchloe dactyloides	0.5	7.8	—	—	—	—
Bromus japonicus	4.0	32.7	11.9	26.6	82.3	90.0
Total grasses	87.6	78.3	89.0	89.0	95.6	95.8
Midgrass						
Bare ground	25.3	5.8	5.0	16.5	1.2	4.1
Bouteloua curtipendula	5.1	11.0	21.1	14.9	52.4	14.7
Stipa leucotricha	43.8	49.3	42.4	46.8	12.5	51.0
Buchloe dactyloides	3.1	4.3	2.0	2.3	—	—
Bromus japonicus	0.1	18.1	9.3	10.8	28.6	14.5
Total grasses	72.6	91.2	86.1	78.3	94.6	91.4
Shortgrass						
Bare ground	28.7	25.0	30.9	20.9		
Bouteloua curtipendula	—	—	—	—		
Stipa leucotricha	6.6	2.7	1.0	2.8		
Buchloe dactyloides	57.4	64.9	10.2	57.0		
Bromus japonicus	0.1	2.0	—	0.8		
Total grasses	69.1	71.9	64.6	72.5		

* CGH, continuous grazing, high stocking; CGM, continuous grazing moderate stocking; DR, deferred rotation; HILF, high intensity, low frequency; Ex. 1 and Ex. 2, exclosures protected from grazing.

From Wood and Blackburn (1984). Used by permission.

the highest percentage of shortgrass community, the lowest percentage of midgrass community and the highest percentage of bare ground (Tables 7.6 and 7.7).

Overall, the most obvious changes in plant community composition were noted among the most intensive grazing systems (CGH and HILF). Still, some species were favoured by grazing (e.g. *Buchloe dactyloides*) and others (*Bromus japonicus*) were not. Across the wide range of grazing pressure considered in this study there were no dramatic changes in community composition but rather changes in the importance of individual species. Such data indicate the resistance of this plant community type to grazing pressure and may suggest a long history of coevolution among large grazers and grasses.

7.3.5 Elephants

The effects of animals on succession can be dramatic, especially if the dominant herbivore is large enough to browse, uproot, and overturn mature trees as is the case with elephants (*Loxodonta africana*). When grassland or woodlands are protected from elephants (not an easy task but nevertheless possible by the use of deep ditches) a pronounced

Table 7.8 Changes in species richness per 4225 m^2 plot in grassland and woodland communities from 1967 to 1981 in Murchison Falls National Park, Uganda when protected from elephants and when not protected from elephants

Community and treatment	No. of species		
	1967	1969	1981
Ground and field layers			
Grassland (protected)	45	47	22
Grassland (unprotected)	46	52	37
Woodland (protected)	36	62	45
Woodland (unprotected)	48	55	54
Shrub and tree layers			
Grassland (protected)	18	17	13
Grassland (unprotected)	15	8	8
Woodland (protected)	17	21	16
Woodland (unprotected)	15	16	13

From Smart *et al.* (1985). Used by permission of Elsevier Applied Science Publishers.

Table 7.9 Numbers of individuals per 4225 m² plot of woody plant species arranged by height classes. Experimental plots were located in Murchison National Park, Uganda and were either protected or unprotected from elephants. The number of species represented by each number is given in parentheses

Community and treatment	No. of individuals		
	1967	1969	1981
< 2 m height			
Grassland (protected)	177 (18)	102 (15)	187 (13)
Grassland (unprotected)	151 (15)	120 (8)	103 (8)
Woodland (protected)	404 (17)	316 (20)	147 (15)
Woodland (unprotected)	276 (14)	362 (16)	218 (13)
2–5 m height			
Grassland (protected)	7 (4)	43 (12)	77 (13)
Grassland (unprotected)	—	—	7 (1)
Woodland (protected)	18 (1)	112 (11)	119 (12)
Woodland (unprotected)	13 (1)	9 (2)	20 (1)
> 5 m height			
Grassland (protected)	—	—	41 (2)
Grassland (unprotected)	—	—	1 (1)
Woodland (protected)	—	8 (1)	38 (4)
Woodland (unprotected)	5 (2)	1 (1)	5 (1)

From Smart et al. (1985). Used by permission of Elsevier Applied Science Publishers.

invasion of trees occurs and palatable grass species decline in importance. For example, Table 7.8 shows changes in a grassland community in Murchison Falls National Park. The tree *Acacia sieberiana* was an important invader as were shrubs such as *Acalypha bipartita* and *Achyranthes aspera*. Species richness of the ground layer decreased under protection from grazing due to the loss of palatable grass species such as *Sporobolus robustus* and *S. pyramidalis*. A less severe decrease in species richness occurred in the grassland community that was unprotected and woody plant invasion was not as rapid (Table 7.8).

In the woodland community protected from elephants, species richness of the ground layer was not severely impoverished (Table 7.8), but again abundant tree regeneration (stems < 2m) was evident (Table 7.9). Woody plant regeneration was found primarily among the dominant tree *Combretum binderianum* but *Acacia* also contributed to the tree stem increase.

Although overgrazing by elephants can impoverish these African plant communities, protection from grazing can also lead to communities of low diversity and low value to grazers, especially in the case of grasslands where palatable grasses are replaced by *Acacia* woodland. Thus, proper management must involve controlled elephant grazing coupled with a working knowledge of how fire interacts with the effects of elephants (Smart *et al.*, 1985).

7.4 SEED DISPERSAL BY ANIMALS

In addition to animals' effects on succession as a result of herbivory, trampling or dunging, there is a clear role of animals in augmenting seed availability. Although a large number of different animal species may disperse seeds, more is known about the activities of birds than any other taxonomic group (Willson, 1986). Accordingly, I will concentrate on a few studies examining the role of birds as seed dispersers during succession primarily to demonstrate that succession managers could utilize locally abundant bird populations as tools in succession management.

Plant species with seeds that are dispersed by birds in eastern North America tend to be over-represented among shrubs and vines with autumn-ripening red or black fleshy fruits (Willson, 1986). The period of highest fruit abundance coincides with the time when migrating bird populations are at a seasonal high (Thompson and Willson, 1979).

Dispersal of seeds by birds can affect the rate and trajectory of succession. Debussche *et al.* (1982) showed that plant species dispersed by frugivorous birds tended to increase in importance with successional age of abandoned orchards in France. Seeds were carried from the orchard edges by birds attracted to senescent fruit trees in the orchard interior. Once birds landed in the perch trees, seeds were dropped or evacuated (regurgitation or defaecation), and this produced a characteristic colonization pattern in relation to the perch trees.

Of particular interest to succession managers is a series of studies (McDonnell and Stiles, 1983; McDonnell, 1986) in which rates of seed deposition by birds were artificially manipulated by placing perches in New Jersey old fields. McDonnell and Stiles (1983) noted that seed input to a 13-year-old field was higher than to a 3-year-old field. The difference in seed input was attributed to a lack of structural complexity in the younger site; this hypothesis was tested by placing

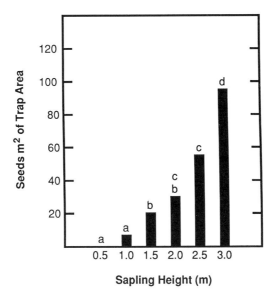

Figure 7.12 Deposition of bird-dispersed seeds under cut saplings of white ash trees in a New Jersey old field. Sapling heights were varied by placing saplings on stakes of different lengths. From McDonnell (1986). Used by permission.

artificial perches of various types in the 3-year-old field. Overall, there were no significant differences among the various perch types; however, seed input was significantly higher in the site that was artificially made to be more complex. A later study (McDonnell, 1986) showed that cut white-ash stems of different height classes, when propped up in old fields differed in their abilities to attract seed-carrying birds (Figure 7.12). More seeds were collected beneath the taller perches (McDonnell, 1986).

It is possible that artificial perches could speed succession in a variety of plant community types. Such a technique would be useful only if speeding succession to a vine- or shrub-dominated phase was desired. Conversely, the observation that perches serve as foci for seed deposition suggests that removal of structures serving as perches would slow succession.

SUMMARY

Succession management can be viewed in two ways with regard to animals. Succession can be manipulated to encourage animal populations, or animal populations can be used to manipulate succession. When succession is managed for animal populations, it is clear that some animal species require early successional phases, some require late successional phases, and still others require a mixture.

During old-field succession, bird species diversity increases primarily as a result of changes in the structural complexity of habitat and as a result of changes in food availability. Logging and subsequent forest regeneration may be detrimental to bird species requiring late successional phases (e.g. cavity nesters), but species characteristic of regenerating shrubby communities may increase. Again, changes in habitat complexity because of disturbance and succession appear to control the suitability of habitat for birds.

Populations of large herbivores such as white-tailed deer and moose can be encouraged by setting back succession. Burning, herbicide spraying, and cutting are all methods that have been used to modify succession and increase the availability of browse for these animals. The spotted owl and the marten are animal species requiring large tracts of mature habitat. Management of succession for these species tends to concentrate on protecting habitat from disturbance.

Grazing and browsing animals modify successional pathways through primary activities such as the clipping of plant parts, trampling of plants, disturbance of the soil, defaecation, and seed dispersal. The effect of animals on succession and the cover of bare ground appear to increase with the intensity of animal activity. In most plant communities a variety of responses to grazing can be observed among the extant plant species. Grazing systems allow succession managers to vary the extent and intensity of animal activity.

Rabbit grazing, when intense, can reduce forb prominence and increase bare ground. Grazing by sheep has been used as a management technique to increase the species diversity of various communities. Cattle grazing is known to increase the invasion rate of shrubs in some plant communities. The resistance of plant communities to change even under strong grazing pressure suggests a long history of plant/grazer interaction.

Artificial perches have potential as a means of speeding succession in some plant communities. Birds are attracted to these perches and deposit seeds from the surrounding plant communities.

8
A landscape perspective

8.1 INTRODUCTION

Previous chapters of this book have dealt with the effects of management activities on plant and animal populations. Although it is clear that mechanical facets of succession management must operate at this level of biological organization, the landscape as a whole must also be considered when management decisions are made for all but the smallest parcels of land. In short, it is difficult if not futile to treat a management unit of vegetation without considering the position of that unit in a larger landscape, without considering interactions among the management units and surrounding landscape elements, and without considering ecosystem-level processes within the management unit. Wildlife biologists have long taken a landscape perspective in their management plans – to a certain extent necessitated by animal mobility – while plant ecologists only recently defined the discipline of landscape ecology as it relates to plant community management (Forman and Godron, 1986).

This chapter will examine the complicating factors contributing to a landscape perspective in succession management. Nature reserves will be used to demonstrate concepts, but this information can also be applied to other natural-resource management situations such as rights-of-way, reclaimed mines, agricultural ecosystems, and lumber-producing forests.

8.2 SUCCESSION MANAGEMENT IN NATURE RESERVES

Nature reserves offer the best example of landscapes where succession can be manipulated to achieve a variety of goals. Some would argue that nature reserves are the last refuges from human disturbance, but this is largely a matter of degree based on the visual presence or absence of human disturbance. Human influence, whether planned and beneficial or unplanned and detrimental, is measurable in nearly every plant community on earth. Indeed, in nature reserves many unique plant communities that harbour a diverse assemblage of plant species are the culmination of succession following many years of

180 A landscape perspective

repeated human disturbance. This is especially true in the heavily populated areas of Europe where heathlands are a prime example. Complete protection from human disturbance in these anthropic communities typically leads to impoverishment as a result of succession (Westhoff, 1971; Medwecka-Kornas, 1977). Thus, active management of nature reserves should not be considered undesirable but rather a force for better resource utilization and preservation.

In nature preservation it is often assumed that, once the reserve is protected, then stability of the biotic community is assured. This assumption ignores the tendency of all natural systems to change through time as a result of internally or externally driven processes (Dolan *et al.*, 1978). What is needed are guidelines for nature preservation that include and recognize the dynamic nature of living systems.

White and Bratton (1980) produced a complete list of prescriptions for managing nature reserves that specifically address this problem. The prescriptions are as follows:

1. Establish aims and goals of preservation.
2. Identify and record the natural resources.
3. Prioritize these resources.
4. Set policy regarding natural disturbance, succession, community restoration, native versus introduced species, rare species, human impact, and changes outside the park boundaries.
5. Begin monitoring natural resources through time.
6. Initiate and fund applied research.
7. Review management plans periodically.
8. Initiate active management where appropriate.

White and Bratton's suggestions frame a cogent statement for a landscape perspective in succession management. Both internal (within reserve) and external (outside reserve) factors must be included in the comprehensive management plan.

8.3 THE INTERNAL DYNAMIC LANDSCAPE OF A NATURE RESERVE

Within the boundaries of a nature reserve a number of succession management schemes can be proposed and a number of landscape elements are relevant. Managers must recognize first that a landscape is an assemblage of patches and corridors embedded in a matrix (Forman and Godron, 1986). Management within nature reserves will

The internal dynamic landscape of a nature reserve 181

concentrate at the patch level. Three types of patches are relevant: remnant patches, disturbance patches, and management unit patches.

Remnant patches are islands of species that differ from the surrounding matrix as a result of a relative protection from disturbance. A single plant community in a nature reserve surrounded by human development would be considered a remnant patch because it differs from the surrounding matrix. Disturbance patches are discrete areas existing as a result of a natural disturbance event. A treefall gap within a forest community would be considered a disturbance patch. Management unit patches are discrete areas of land that receive a single management treatment to modify succession in that patch. From Chapter 1, management unit patches can be subjected to designed disturbance, controlled colonization, or controlled species performance. By maintaining the boundaries of remnant patches or by creating new management unit patches, the following succession management schemes are possible for a hypothetical three-patch nature reserve (Figure 8.1). This range of management schemes is by no means complete but it will serve to demonstrate how succession within the framework of patches can be manipulated to achieve a variety of land use goals.

While it is clear that successional pathways can be changed using techniques described in Chapters 4–7, in all of the following schemes only the mix of successional states is changed rather than the direction of succession.

8.3.1 Complete protection

Under complete protection all major person-generated disturbances are eliminated or controlled (Figure 8.1). This includes disturbances used to manage succession. With complete protection the three remnant patches will converge toward a single remnant patch. Overall, landscape diversity will decrease through time as will species diversity. The entire reserve would be considered a single management unit patch. This approach is appropriate if the goal is to maximize populations of forest-dwelling bird species over the long term.

8.3.2 Complete disturbance

With complete disturbance the entire reserve is treated as a single management unit patch from the outset. The forest and 15-year-old field are cleared of woody vegetation and then the entire reserve tilled to set back succession to a bare ground stage. Secondary succession would probably differ among the sites where the three remnant

182 A landscape perspective

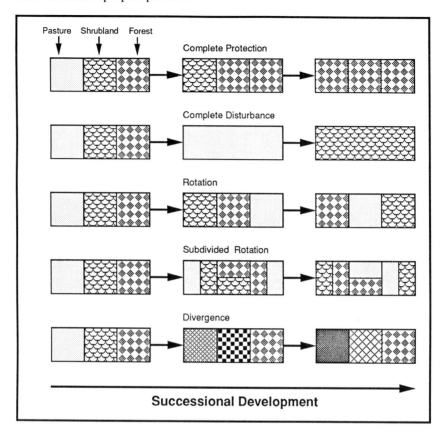

Figure 8.1 Schemes for developing management unit patches of different ages, sizes and distributions in a hypothetical nature reserve that originally consists of three remnant patches.

patches originally existed as a result of differences in propagule banks (Pineda *et al.*, 1987), but overall landscape diversity and species diversity would decrease. Three remnant patches would be converted into a single remnant patch. This approach is appropriate if the goal is to maximize populations of game birds that require a large block of early successional habitat.

8.3.3 Rotation

Under a rotation management plan the original remnant patch boundaries remain intact, but all remnant patches are converted into management unit patches. Management activities strive to maintain three distinct successional phases (e.g. 0–5-year-old field, 25–30-year-old field, and 50–55-year forest) separated by an arbitrary number of years to maximize between patch differences. Overall landscape diversity would remain unchanged. Trends in species diversity depend on species' responses to repeated management activities. Rotation is appropriate if the goal is to maintain a mixture of plant communities as existed at the time of reserve acquisition.

8.3.4 Subdivided rotation

In subdivided rotation each remnant patch is subdivided into smaller management unit patches. Within each original remnant patch a rotation schedule is set up to maintain different successional phases at the subunit level. The overall effect is to increase the number of management unit patches but patch size decreases and edge area increases. Subdivided rotation is appropriate if the goal is to maximize habitat diversity as well as edge habitats.

8.3.5 Divergence

In divergence it is assumed that the three remnant patches in the reserve do not represent a predictable sequence of succession. Rather, they are simply plant communities that may change, but the time frame of this change is so long that it does not warrant careful attention. The outcomes of these trajectories, whatever they may be, are considered desirable. Moreover, since people strongly influenced the development of these communities up to the present point, then continued human influences (e.g., introduced species, grazing, vehicle traffic) should not be regulated. Management is largely non-existent. Indeed, many nature reserves that are not managed or are not well-controlled with regard to human disturbance will default to this type of management plan.

8.4 OPTIMUM MANAGEMENT UNIT PATCH SIZE

A fundamental question regarding the proper succession management scheme for a nature reserve deals with upper and lower limits of management unit patch size to maintain viable plant and animal populations through time. With the development of the theory of island biogeography, an extensive literature emerged that attempts to determine the minimum critical size for nature reserves using species–area relationships, extinction rates, and immigration rates as guiding

184 *A landscape perspective*

Figure 8.2 (a) Cantilevered edge maintained by mowing to the bases of trees. (b) Dripline edge occurs when mowing is repeated at the canopy dripline. (c) Advancing edge that develops when mowing ceases.

principles (Diamond and May, 1976). Although island biogeography has made important contributions to theoretical considerations of size and shape of nature reserves, this theory has not been well tested in a variety of ecosystems and should not be viewed as a hard-and-fast body of knowledge for making resource management decisions (Simberloff, 1986). Indeed, it is assumed that management activities, at least in heavily populated areas of the world, are more important as determinants of plant and community structure than are processes associated with island biogeography (Gibson, 1986). Moreover, the design of nature reserves with regard to size and shape is often based on considerations other than species–area relationships (Game and Peterken, 1984). Among these less theoretical but nonetheless real factors are historical boundaries, monetary resources, land area available for acquisition, uncontrolled natural disturbances, and topographic limitations. Finally, the theory of island biogeography is not well integrated with current theories of succession.

A more workable solution for assessing the proper size of management unit patches in succession management is to examine

186 A landscape perspective

relationships among edge area, interior area, and disturbance patch area as management unit patch area is varied. These relationships are relevant to interactions among individual management unit patches but also have important bearings on the design of entire nature reserves.

Edge is created when two different landscape elements abut. The edge concept can be traced to an earlier ecological concept known as the ecotone, a zone where one plant community merges into another. Ecotones typically exist as a result of species assemblages varying across the landscape in response to environmental factors. Ecotones can be gradual or abrupt depending on species responses and the rate of change of the environmental gradient (Whittaker, 1975). Edges, on the other hand, are created by human-generated environmental gradients. The edge of a forest patch abutting a pasture would not exist without (1) a human decision to preserve the forest patch and (2) a human decision to create and maintain the pasture by grazing.

Depending on management practice and patch age, edges can take a variety of forms. Ranney *et al.* (1981) identified canopy dripline, cantilevered, and advancing edges at the outside of forest islands (Figure 8.2). Dripline edges are maintained by vegetation removal to the canopy dripline; cantilevered edges are maintained by plant removal to the tree trunk base; advancing edges may develop in the periodic absence of management.

Edge habitat has been studied from a variety of perspectives. Edges of forests differ significantly from forest interiors with regard to microclimate and biotic communities. Ranney *et al.* (1981), in a study of forest islands in Wisconsin, noted the following differences between forest edges (15 m strip) and forest interiors: edges supported higher tree basal area and more tree species than interiors; edges supported more tree species with high light requirements than interiors. Even successional rates and trajectories of edges and interiors differ. Edges reach equilibrium faster than interiors, but edges at equilibrium support plant species assemblages that are not representative of old-growth status. Edge communities tend to resemble the plant communities developing in disturbance patches of interior forest communities (Ranney *et al.*, 1981).

With increasing forest fragmentation, edge habitats assume greater importance in the successional development of forest communities. For example, Whitney and Runkle (1981) compared forest communities of the edges and interiors of two forest patches in Ohio, one patch a second growth forest and the other an undisturbed forest. They found that edge versus interior differences within a patch tended

Optimum management unit patch size 187

Figure 8.3 A small patch of forest in a pasture that is dominated entirely by plant species typical of the forest edge or disturbed sites. This patch, well isolated from other patches, does not support species typical of forest interiors.

to be more pronounced than age differences of the interiors of the two forest patches.

The forest edge serves as an important refuge for shade-intolerant species, and these species in turn provide propagules for other edges and disturbance patches (McDonnell and Stiles, 1983; Debussche *et al.*, 1982). However, when patch size gets increasingly small (Figure 8.3), edge species dominate a forest island and may eventually usurp shade-tolerant tree species. Indeed, Levenson (1981) found that patches less than 2.3 ha in area tended to function entirely as edges. For circular or square islands he noted that interior area and edge area change at different rates as patch size is varied; a critical point is reached where interior is no longer a viable habitat (Figure 8.4).

The form of a forest edge can also influence colonization of plants into adjacent openings. An interesting study by Hardt and Forman (1989) dealing with colonization of surface mines in eastern US showed that concave forest boundaries were more conducive to the

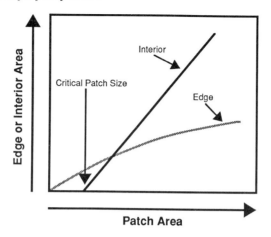

Figure 8.4 The relationship between edge area or interior area as total patch area increases. This relationship holds approximately for circular or square patches. From Levenson (1981). Used by permission of Springer Verlag Publishers.

movement of tree species into adjacent surface mines than were straight or convex boundaries. Presumably, concave boundaries tended to aggregate the stems of clonal trees; this in turn encouraged the activities of seed-carrying animals.

A number of studies have stressed the importance of preserving large forest islands, but, depending on target species, this also is not a general rule. Levenson (1981) maintained that conservation of large forest patches (> 4 ha) should be given priority because these support self-perpetuating populations of shade-tolerant tree species. Forman *et al.* (1976) came to a similar conclusion because of trends in bird diversity relative to patch size. However, Game and Peterken (1984) found with woodland herb species that a large number of small reserves would preserve a greater number of species than a few large reserves. Rosenberg and Raphael (1986) predicted that old-growth forest patches less than 20 ha in the Pacific Northwest would begin losing species characteristic of this community type.

A final consideration regarding the size of management unit patches is the natural disturbance regimen of a nature reserve. Natural disturbances, like human-generated disturbances, tend to operate in a patchy fashion through time (White, 1979). Patch dynamics driven by natural disturbances will be superimposed on any managed succession pathway (Figure 8.5). In the long-term development of forest stands such patch processes are important determinants of ecosystem structure and function (Bormann and Likens, 1979). Pickett and

Distribution and connection 189

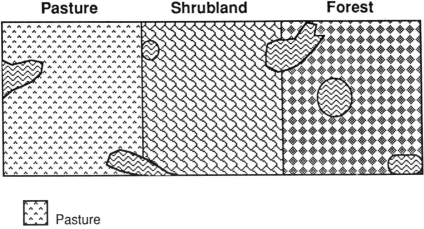

Figure 8.5 Distribution of disturbance patches in a hypothetical nature reserve. All disturbance patches presented here are smaller than the largest management unit patch. However, a single disturbance patch could encompass the entire nature reserve.

Thompson (1978) suggested that patch processes driven by natural disturbance should be an integral component of natural resource management decision-making. Specifically, they made a case for managed land areas large enough to include the largest disturbance patch, different ages of disturbance patches and habitats to provide propagules for recolonization of disturbance patches. Before a minimum management unit patch size is chosen, an assessment must be made of the type, distribution, size, and frequency of natural disturbances.

8.5 DISTRIBUTION AND CONNECTION

In addition to decisions regarding management unit patch size, resource managers must also arrive at decisions regarding the distribution of these patches within the boundaries of a nature reserve. The distribution pattern has consequences primarily for animal dispersal and migration, but may also be important for the movement and availability of plant propagules because some plant propagules move with animals.

If management unit patches are contiguous and animals readily utilize them as habitat, resource managers can then be somewhat lax in the choice of distribution pattern. Often, however, management unit patches interact in a complex manner and careful consideration must be given to juxtaposition and connection (Rafe *et al.*, 1985; Noss and Harris, 1986). Management unit patches can be created that are not conducive to animal movement (e.g. tilled land, pastures, and clear-cuts). This may in turn insulate other patches from animal and plant populations. For example, Soutiere (1979) found that martens in Maine did not utilize and would not cross 0–15-year-old patches of regenerating forest created by clear-cutting. Warner and Joselyn (1986) found that roadside habitat developed for ring-necked pheasants was utilized at a higher rate when roadsides were adjacent to hay and small-grain fields. Levenson (1981) noted that American beech (*Fagus grandifolia*) was being eliminated from forest islands in Wisconsin because of island isolation and an inherently low dispersal capacity of the species. It may be necessary to connect a system of management unit patches via corridors (Figure 8.6) or the problem of connectivity may be solved by planning management unit patches so that interactions are maximized.

Linking management unit patches with corridors can be accomplished with a variety of landscape elements. Forman and Godron (1986) identified line corridors, strip corridors, and stream corridors. Such corridors, depending on their structure, can serve as habitats, conduits, barriers, or modifiers of the environment (Forman and Godron, 1986).

Line corridors such as hedgerows do not have interiors distinct from edges. Strip corridors such as power line corridors are wide enough so that the interior is distinct from the edge. Twelve metres appears to be the critical width separating line corridors from strip corridors at least as indicated by herbaceous plant species (Forman and Godron, 1986). Stream corridors include the stream proper as well as the vegetation growing along the streambank. The best corridors for animal movement support vegetation higher than the surrounding matrix, but a landscape element as simple as a fencerow will serve as a route for the movement of animals from one patch to another (Wegner and Merriam, 1979). The connection of management unit patches via corridors may necessitate succession management in the corridors since even these strips of vegetation will undergo successional change through time (Schulze *et al.*, 1986).

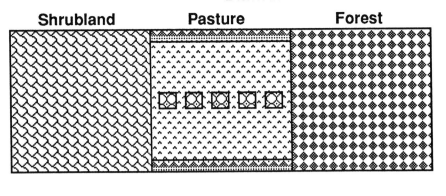

Figure 8.6 Two methods of connecting the shrubland and forest management unit patches. Migration routes can be continuous corridors such as hedges or forests at the edges of the pasture. Patches of tall vegetation can also be distributed as stepping stones.

Connections among management unit patches may also be maintained with small islands of suitable habitat rather than continuous corridors. Forman *et al.* (1976) stressed the importance of small forest islands as stepping stones for birds as they moved among larger forest islands in New Jersey. Management unit patches with low open vegetation can be used to discourage animal movement. Verme (1965) suggested that adjacent patches of white-cedar (*Thuja occidentalis*) managed for white-tailed deer in Michigan should be large enough so that deer would not use the clear-cuts and thus inhibit white-cedar regeneration after cutting. Therefore, patch size, patch distribution, and corridors can be changed in such a manner to either encourage or discourage animal use depending on succession management goals.

8.6 EXTERNAL FACTORS

Even the most complete succession management plans are subject to interference from a number of external factors. Some of these external factors are controllable whereas others are nearly impossible to control. The following sections will examine some of the external factors that can potentially change successional pathways within a nature reserve. Emphasis will be given to those factors that are directly or indirectly the result of human activities, since these are most commonly associated with resource management problems.

8.6.1 Invasive and introduced species

Introduced species pose a formidable problem for those individuals involved in succession management of nature reserves. This is especially true when a specific management goal is to preserve species of the native flora and fauna. Introduced species, whether plants or animals, lead to changes in successional pathways as well as extinctions among the local native species.

Few nature reserves are immune from introduced species. However, Usher (1988) showed that island nature reserves are more likely to be invaded by introduced species than are continental savanna or woodland reserves. In addition, he determined that there is a negative relationship between reserve area and the number of introduced species, at least for nature reserves in arid and Mediterranean climates. Clearly, introduced species can be a severe problem in very small nature reserves where the invasion can quickly extend throughout the entire protected area.

Few descriptive studies are available where changes in successional pathways have been documented as a result of a biological invasion. Rather, the state of knowledge with regard to introduced species is still at a stage where resource managers are attempting to determine the scope of the problem (Usher, 1988). In many cases where the presence of introduced species is well documented, little is known about the life histories of the introduced species beyond simple feeding preferences or single-species interactions (Brockie *et al.*, 1988).

Regardless of the lack of data on introduced species and their potential effects on succession, it can be assumed that introduced plants and animals with certain characteristics will cause the largest impact. Among animals, the most severe changes in succession will result from large grazers or browsers that tend to reduce palatable native plants and also disturb the soil by trampling or rooting. Feral goats (*Capra hircus*) on a number of island nature reserves are a good example of an introduced species that appears strongly to modify the successional development of native plant communities (Mueller-Dombois, 1981; Brockie *et al.*, 1988). The European wild boar (*Sus scrofa*) damages the understorey of beech forests in the Great Smoky Mountains National Park by rooting (Bratton, 1975). Repeated hog rooting leads to reduction in cover and number of understorey species.

Among introduced plant species, greatest effects on successional pathways can be expected from trees, shrubs, and vines. These species have the greatest potential for competing with native plants and also for strongly modifying the microenvironment. However, modifications of successional pathways by introduced species are not limited solely to these growth forms as evidenced by the large number of exotic annual and perennial grass and herb species that now dominate the weed floras of agricultural ecosystems.

Invasion and spread of introduced plant species are favoured by human-generated disturbance (Fox and Fox, 1986), a fact that should instil caution in all managers who implement management activities without prior testing. Even the presence of humans as indicated by visitation rates to nature reserves, can be correlated with the numbers of introduced species in nature reserves (Usher, 1988). In Australian nature reserves, exotic plant species readily invaded where fire frequency or soil nutrient levels were increased near urban settlements (Bridgewater and Backshall, 1981; Clements, 1983). Likewise, frequent fires and intense grazing by livestock facilitated the invasion of South African nature reserves by a number of aggressive woody plant species (Macdonald *et al.*, 1988). Ehrenfeld (1983) found that pine barren wetlands in New Jersey adjacent to human-disturbed or developed sites lost representatives of the native shrub flora and became dominated by introduced herb and vine species. This change in community trajectory was linked to increases in nutrient availability as a result of runoff from developed areas.

A primary problem associated with the invasion of introduced plant species is a reduction in species and habitat diversity. Some introduced plant species form monospecific or nearly monospecific stands. In northern Kentucky the introduced shrub Amur honeysuckle (*Lonicera maackii*) readily invades open sites and early successional forests near urban developments. Production of a thick and impenetrable shrub stratum (Luken, 1988) leads to complete elimination of understorey species and tree seedlings on some sites. Amur honeysuckle is long lived and it is likely that successional pathways in open sites (Luken and Thieret, 1987) and in forests (Luken, unpublished data) are being

194 *A landscape perspective*

modified. The exact direction of these changes is, however, not known due to the short period of time this shrub has been a part of the regional flora.

Because resource managers may be faced with the presence of many introduced species there must clearly be a prioritization process to determine which introduced species are causing the greatest impact (Usher, 1988). Once a species is targeted for control (Figure 8.7), a variety of options is available such as manual or mechanical removal, fencing, pesticide use, or biological control. The success of an eradication programme will depend on the extent of the problem and on the availability of monetary resources. In the Cape of Good Hope Nature Reserve in South Africa 39% of the reserve management budget is devoted to control of introduced woody plant species (Macdonald et al., 1988). In addition, control programmes should be of a long-term nature to determine if reinvasion occurs and also to determine how native plant communities respond to the elimination of an introduced species.

It is possible that introduced species in some nature reserves may be so widespread and pervasive that nothing can be done to reverse the trend. In such a situation it is perhaps best to study the invasion thoroughly and learn as much as possible so that the information can be used to prevent such an event in other less-polluted nature reserves.

8.6.2 Off-road vehicles

Off-road vehicle (ORV) traffic, whether controlled or uncontrolled, is an external environmental factor that can strongly modify the development of plant communities on nature reserves. The majority of studies dealing with the effects of ORVs on succession are done in arid regions. The reasons for this are twofold. First, arid regions, at least in the US receive much ORV pressure. Second, vegetation change in arid regions is slow and there is concern that ORV traffic could cause long-lasting effects. In arid regions the primary ORVs are motorcycles or

---

Figure 8.7 Chemical control of prickly-pear (*Opuntia ficus-indica*) in the Andries Vosloo Nature Reserve, South Africa. Much of the infestation was controlled by introduction of two insect herbivores (*Cactoblastis cactorum* and *Dactylopius opuntia*), but remaining plants require chemical treatment. (a) A control team injects cladodes with MSMA (Monosodium acid methanearsonate). (b) A prickly-pear individual killed by MSMA injection. Treated areas are slowly recolonized by native plants. Photographs courtesy of Dr I.A.W. Macdonald.

trucks; in other plant communities, snowmobiles are a common disturbance.

In deserts, the direct effects of ORVs include damage or destruction of existing plants as well as soil compaction. Reduction in perennial plant cover occurs in areas of ORV use, and this reduction is dependent on the intensity of ORV use (Hall, 1980; Bury and Luckenbach, 1983). Some perennial plants resprout after damage if the assaults are not repeated frequently (Vollmer *et al.*, 1976). However, over the long term it is soil compaction that appears to have the most significant effects on succession when ORV traffic ceases.

In general, four-wheel-drive vehicles have a greater impact than motorcycles; wet soils are compacted more readily than dry soils (Adams and Endo, 1980; Webb *et al.*, 1983). Studies of old military camps in the Mojave Desert show that soil compaction caused by vehicle traffic can affect plant communities as long as 40 years after the original disturbance (Lathrop, 1983; Prose *et al.*, 1987). With increasing soil compaction, dominants of the original plant community such as the long-lived shrub (*Larrea tridentata*) do not readily recolonize sites, instead, pioneer shrubs such as *Ambrosia dumosa* and *Hymenoclea salsola* assume greater importance (Prose *et al.*, 1987).

Snowmobiles are also a problem in other plant communities primarily as a result of the snow compaction that they cause and the associated changes in microclimate. Changes in percentage cover of herbaceous species have been noted in snowmobile tracks (Keddy *et al.*, 1979) suggesting that snowmobile traffic should be limited in areas supporting rare or endangered species.

8.6.3 Landscape changes outside nature reserves

Boundaries of nature reserves drawn by people often do not represent boundaries for the transfer of materials among landscape elements. Dramatic changes in ecosystems outside a nature reserve may lead to changes in successional pathways inside nature reserves. For example, Schroeder *et al.* (1976) documented changes in the succession rate on barrier islands in North Carolina as a result of construction of barrier dunes. Succession on the islands was subsequently accelerated to a phase dominated by woody plant species due to a reduction in overwash (Schroeder *et al.*, 1976). Bakker *et al.* (1987) documented how the attempts of succession managers to create species-rich grasslands in The Netherlands have been thwarted by nearby agricultural practices. The construction of a water diversion ditch at the boundary of the reserve caused an increase in flooding of low

External factors 197

elevation communities with eutrophic waters. As a result, in the low elevation communities, plant species with high nutrient requirements such as *Glyceria maxima* and *Carex acuta* spread at the expense of less productive species. Dams that change the fluvial dynamics of waterways can initiate new successional pathways (Figure 8.8) and also allow succession to proceed beyond what would normally occur under a seasonal flooding regimen (Bravard *et al.*, 1986).

Although it is difficult and sometimes impossible to foresee the impact of human disturbance outside a nature reserve, it is nevertheless prudent when acquiring land for a nature reserve to acquire entire watersheds or to attempt regulation of human disturbance outside the reserve that may affect succession inside the reserve.

8.6.4 Air pollutants

Air pollutants readily cross the boundaries of nature reserves and can have either detrimental or stimulatory effects on plants. Many air pollutants ranging from dust to complex organic molecules travel long

Figure 8.8 Succession on the banks of the Lower Ume River in Sweden after the water level was permanently lowered. Scots pine (*Pinus sylvestris*) has invaded what was once a willow zone. Photograph courtesy of Dr C. Nilsson.

distances and are deposited in virtually every ecosystem on earth. The effects of these air pollutants on ecological succession are largely unknown. In some cases the pollutants are causing the senescence of plant populations (Smith, 1981), but the complexity of the die-back process and the large number of pollutants available are making it difficult to demonstrate cause and effect, let alone impacts on long-term community dynamics. Modelling efforts have been made to predict long-term community changes as a result of various air pollutants (West *et al.*, 1980). These exercises suggest that relatively small decreases in growth of individual species can lead to novel species replacements over the long term.

A series of studies carried out in The Netherlands demonstrated that atmospheric ammonia released from manure spread on agricultural fields could indeed be taken up by plants (Heil *et al.*, 1988). Bobbink and Willems (1987) suggested that ammonia air pollution is a primary reason why the aggressive grass *Brachypodium pinnatum* is achieving dominance in a wide range of chalk grassland and heathland reserves in western Euope (Figure 8.9).

Although there is little direct action a resource manager can take to stop air pollutants from entering a nature reserve, it is important that monitoring programmes be set up to assess reductions or increases in plant populations coupled with measurements of common air pollutants. Such data could be used to address regulatory agencies that set standards for emissions if and when a trend is detected.

SUMMARY

Nature reserves are landscapes that are afforded various degrees of protection. They are in many cases islands in a matrix of human-disturbed land. In any reserve, policies need to be established regarding plant communities or species of most value, natural disturbances, human-generated disturbance, and succession management. Both internal and external factors can affect succession within nature reserves, and succession management plans must take these factors into consideration.

The landscape perspective in succession management recognizes various landscape elements such as disturbance patches, remnant patches, management unit patches, and corridors. These landscape elements interact by virtue of patch size and patch distribution. Management unit patches should be large enough to contain disturbance patches and so that edge effects do not dominate the entire patch. Individual management unit patches should be connected by

Figure 8.9 Successional changes in heathlands of The Netherlands from 1979 to 1982 as a result of atmospheric nitrogen input. (a) Heathland in 1979 dominated by *Calluna vulgaris*. (b) The same heathland in 1982 now dominated by the grasses *Molinia caerulea* and *Deschampsia flexuosa*. Photographs courtesy of Dr G.W. Heil.

small islands or corridors so that living organisms and propagules will readily move among patches.

External factors that may affect succession within nature reserves include introduced species, large-scale landscape modifications, off-road-vehicle traffic, and air pollutants. Some of the factors are controllable while others are not. In all cases monitoring programmes should be in place to determine how a specific factor is affecting succession.

9

Information systems for prediction and decision-making

9.1 INTRODUCTION

In addition to the myriad of tools and techniques for succession management available to resource managers (Chapters 4–7), there are a number of information systems that aid decision-making. These information systems, many of which rely on readily available computer software and hardware, provide resource managers with several types of output.

This chapter will first consider succession models. Such models allow one to predict the outcome of succession many years in the future, but most importantly the response of succession to various management practices can be simulated. Succession models originally developed in the 1970s and further refined during the last decade have reached a high level of sophistication. They are now used in many different plant communities but are particularly well suited to forests (Shugart, 1984). Expert systems treated in this chapter are not yet well-developed enough to support succession management, but they have great potential for simplifying the decision-making process. And lastly, resource management algorithms will be examined to demonstrate how a great variety of data, including succession data, can be incorporated into decision-making processes to achieve complex natural resource management goals.

9.2 SUCCESSION MODELS

Shugart (1984) provided an in-depth analysis of various types of forest succession models. In general, models of interest to succession managers are those that predict community-level or landscape-level changes through time as a result of management or natural disturbance.

Gap models, a special group of succession models, simulate plant community changes in small patches of land. They are driven by the knowledge that establishment, growth, death, and replacement of individual trees can be described by a system of equations that is deterministic or stochastic in function. As one example consider the FORET model developed by Shugart and West (1977) to simulate succession in Tennessee deciduous forests. This model included subroutines for estimation of the following parameters:

1. Tree growth – estimated as a function of climate, leaf area, crowding, and tree size.
2. Tree birth – estimated as a function of seedbed conditions, climate, and herbivory.
3. Tree sprouting – estimated as a function of tree size and sprouting ability.
4. Tree death – estimated as a function of maximum tree age.

Shugart and West (1977) included 33 tree species in this model and calculated successional changes within a 1/12 ha plot. As a validation procedure of the model they modified the species present and simulated effects of the chestnut blight, a disease caused by a parasitic fungus that eliminated nearly all American chestnut (*Castanea dentata*) throughout eastern North America. One run of the model was done with chestnut present, and another was done with chestnut deleted.

This model with chestnut absent predicted (Figure 9.1) that stem density would decline rapidly during the first 100 years and then increase with subsequent stabilization. Biomass would increase rapidly to year 40, and then the rate of increase would slow presumably as the stand began to thin. Clear changes in individual species importance were predicted by the model (Figure 9.1). With chestnut present, black oak assumed importance early in stand development and increased in importance through time while yellow poplar maintained high importance throughout stand development. With chestnut eliminated, white oak and chestnut oak assumed greater importance (Figure 9.1).

When stand characteristics of a pre-blight Tennessee forest were compared to the stand characteristics predicted by the model, adequate agreement was noted. Therefore, the model was sufficiently sensitive to changes caused by the deletion of a single but clearly important tree species (Shugart and West, 1977). Unfortunately, the researchers did not provide field data on present-day forests. Thus we do not know how well the model predicted succession changes as a result of chestnut loss.

This FORET model was expanded and transformed into the KIAMBRAM model in an effort to simulate the effects of forest harvest on succession in subtropical rain forest in New South Wales

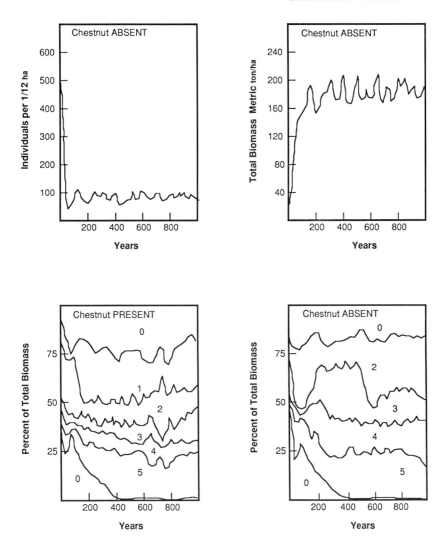

Figure 9.1 Simulated trends of stem density, biomass and species composition for a Tennessee deciduous forest as predicted by the FORET succession model. Species composition was simulated with or without the presence of American chestnut (*Castanea dentata*). To exaggerate trends of important species, less important species are not shown in the species composition figures. The contribution of a species to total biomass is indicated by the size of the numbered area between upper and lower lines. 0, other species, 1, American chestnut (*Castanea dentata*), 2, yellow poplar (*Liriodendron tulipifera*), 3, white oak (*Quercus alba*), 4, chestnut oak (*Quercus prinus*), 5, black oak (*Quercus velutina*). From Shugart and West (1977). Used by permission of Academic Press.

(Shugart *et al.*, 1980). Modification to the model included equations to describe the effects of tree fall gaps, lumbering, pulsed seed production, and strangling vines. The KIAMBRAM model predicted that subtropical forests could sustain clear-cutting every 30 years but the forest composition would change through time so that there was a higher representation of early successional species. Such a prediction is not surprising, but overall it does demonstrate the utility of such models in predicting the long-term outcome of various management practices.

A somewhat different approach to succession modelling was pioneered by Noble and Slatyer (1977) with the identification of vital attributes of different plant species. They concluded that every plant species could be categorized with regard to the method of arrival or persistence of propagules at a site, the conditions under which a species establishes and grows to maturity, the longevity of individuals of a species, and the time it takes to reach critical life history stages. Different methods of persistence include:

1. The production of widely dispersed seeds (symbolized D).
2. The production of long-lasting seeds that accumulate in the soil and may survive disturbance (symbolized S).
3. The production of seeds above ground that are stored in cones or fruits (symbolized C).
4. Vegetative reproduction (symbolized V).

Different conditions for establishment include:

1. Species that establish among established individuals of all other species (symbolized T).
2. Species that establish only after a disturbance eliminates competing individuals (symbolized I).
3. Species that cannot establish immediately after a disturbance but will establish only when other mature individuals are established (symbolized R).

Critical life history events include:

1. The production of sufficient propagules to survive a disturbance (symbolized p).
2. The time it takes for an individual to reach maturity and become established (symbolized m).
3. The senescence and local extinction of a species (symbolized l).
4. The loss of propagules from a site (symbolized e).

The first two vital attributes can be combined in a variety of ways, and the corresponding sequence of life history events can be matched to these (Figure 9.2). For example, with a DT species, one with widely dispersed seeds that readily establishes among other individuals, the availability of propagules (p) is the same as time zero because the propagules are available from a previous generation. Maturity (m) is reached at some later point in time followed by senescence (l) and loss of propagules from the site (e). This can be contrasted with a CT species where maturity (m) must be reached before propagules are available.

Using the example of a *Callitris*–eucalypt forest community in Australia, Noble and Slatyer (1977) demonstrated how this classification can be used to model succession through repeated fires (Figure 9.3). *Callitris* is considered a CT species because it stores seeds above ground, takes 7 years to reach maturity, and establishes among other plants. Eucalypts are VI species because of their ability to regenerate vegetatively and because they establish only after a disturbance removes existing plants. When a site supporting both CT and VI is

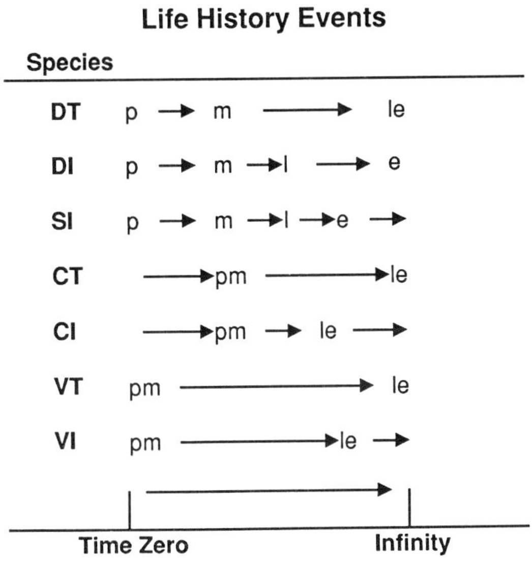

Life History Events

Species	
DT	p → m ⟶ le
DI	p → m →l ⟶ e
SI	p → m →l →e →
CT	⟶pm ⟶le
CI	⟶pm → le ⟶
VT	pm ⟶ le
VI	pm ⟶le →

Time Zero ⟶ Infinity

Figure 9.2 Various combinations of vital attributes (method of persistence and required conditions for establishment) producing a number of hypothetical species and the sequence of important life history events (p, production of propagules; m, time until maturity; l, local extinction, and e, loss of propagules) associated with each combination. From Noble and Slatyer (1977). Used by permission.

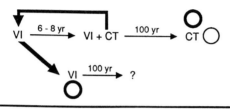

Figure 9.3 (a) Vital attributes of two tree species in an Australia *Callitris*–mallee community and their sequences of life history events. (Mallee is VI because it reproduces vegetatively only after fire whereas *Callitris* is CT because it stores seed above ground in cones and does not require a disturbance to establish.) (b) Predicted successional trends in the absence of fire. (c) Predicted trends with fires at various stages of community development. From Noble and Slatyer (1977). Used by permission.

burned, VI will regenerate immediately. CT seedlings will germinate and eventually codominate with VI. In the absence of fire, VI will die out leaving only CT. If the community dominated by CT is burned, CT will regenerate but VI will not. On the other hand, if the community is burned while CT and VI dominate, it will revert to VI domination. Repeated burning in the VI stage could lead to a self-perpetuating community dominated by VI (Figure 9.3).

The logical extension of forest succession models is to attempt the

Succession models

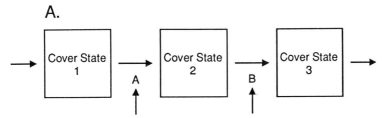

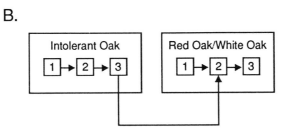

Figure 9.4 (a) The concept of landscape modelling as shown by the transition of one cover state to the next. (b) One potential cover type transition in the Great Lakes region of Michigan. Each large block represents a forest type and the three smaller boxes within represent seedling, pole timber, and saw timber stands, respectively. From Shugart et al. (1973). Reprinted from the Society of American Foresters.

simulation of entire landscapes composed of many successional states. Such an approach first requires that the landscape units or patches be separated into a series of distinct states. Functions for the transition of one state to the next (area flow) must then be determined (Figure 9.4). Shugart et al. (1973) divided forests of the US Great Lakes region into 15 different cover states and then simulated trends for the entire landscape over many years.

Succession models can also be used to help managers maximize animal populations. One of the first such uses of these models for this purpose was presented by Giles and Snyder (1970). They sought to maximize elk populations in the Clearwater Range of western US. Beginning with the basic assumptions that elk populations are limited by available forage, and that forage biomass varies as a result of both succession and management activities, Giles and Snyder (1970) developed a model that was based at the level of management units. A field-based inventory system allowed the researchers to assign a successional curve to each management unit. These curves varied

208 Systems for prediction and decision-making

depending on past management activities, fire history, and slope. By simulating succession across many management units, Giles and Snyder showed how management activities could bring about a gradual increase in forage availability as well as in elk population size (Figure 9.5).

It is now common in the planning of national forest management in the US to link succession models with wildlife habitat suitability models. For example, the Dynamically Analytical Silvicultural Technique (DYNAST) (Boyce, 1977) is a forest succession model that can be used to predict how the mix of habitats in various stages of succession will change through time. This output is then used to determine the species and population sizes that a management unit can support (Benson and Laudenslayer, 1986).

9.2.1 Advantages and limitations of succession models

As with all tools in succession management, there are both advantages and disadvantages to succession models. A model is designed to mimic succession in the real world. Ecologists couple descriptive studies of succession with model development. When such an exercise is undertaken it is in essence an attempt to see whether the important factors controlling plant community change have been adequately

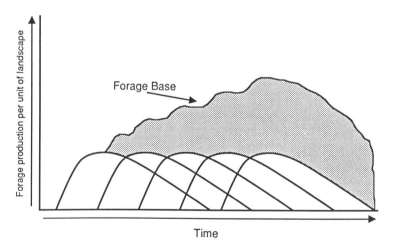

Figure 9.5 A simulated trend in forage production across the landscape as a result of setting a series of management units in staggered successional development. Each curve represents forage production of a single management unit during succession. From Giles and Snyder (1970). Society for Range Management. Used by permission.

identified and measured. As such, models help ecologists to understand what they do or do not yet know about succession. In this sense, models are valuable tools.

Provided the important parameters have been identified, and provided adequate data are available for the important species, models allow resource managers to ask the question: what if? This is indeed a unique advantage to models, especially taking into consideration the relative lack of applied research where long-term effects of management activities are measured directly. Up to now the majority of models directed at answering resource management problems have been largely forest models with applications centred on timber harvesting or the long-term effects of air pollutants (Shugart, 1984). However, it is likely that the future will see more modelling efforts in non-forest communities, primarily with the goal of understanding long-term effects of succession management activities across the landscape.

There are, however, a number of limitations to models that resource managers should take into consideration. The range of parameters in a model is dependent on the researcher's knowledge of the system being simulated and of its individual species. Undoubtedly, in some plant communities the range of parameters now included in existing models is adequate to predict succession, at least within the range of validation procedures used. Models, however, will never be able to include all parameters affecting succession. It is impossible to predict certain events that may dramatically alter successional pathways or rates at some point in the future (e.g., introduced species, climate changes and stochastic disturbance events). This is especially relevant when simulations are projected far into the future. Moreover, some aspects of plant/plant, plant/animal, and plant/landscape interactions are not yet included in models because of a lack of data even though it is understood that these interactions may influence succession.

Succession models are often general enough so that they can be used in a wide variety of situations. Unfortunately, when a model is developed in one system and then used in another system it must be modified. The lack of a single model that can be easily changed to simulate succession in a variety of plant community types is likely to limit model use by resource managers. Most resource managers do not have the expertise to modify existing models, and thus the use of models is contingent on the availability of modellers.

9.3 EXPERT SYSTEMS

Expert systems or consultation systems are tools that resource managers can use to make management decisions (Starfield and

Bleloch, 1983). Although expert systems exist for wildlife management (Marcot, 1986), they are not yet widely available for succession management. It is likely that they will become more popular in the future for the following reasons. They are easy to use and they give straightforward answers to simple questions. This is in direct contrast to succession models where there is complex output and the output is often difficult to interpret regarding the best management scheme. Expert systems are based on a wide variety of information including mathematical, experiential and anecdotal data. Again, this can be contrasted with succession models where the driving functions are almost always mathematical. In many cases the mathematical functions used in succession models are poorly validated, thus giving a false sense of accuracy (Noble, 1987). Lastly, expert systems provide information quickly.

The heart of an expert system is a series of rules. For each management problem or biotic system a different set of rules must be developed. Rules systems are often decided by committee action (Starfield and Bleloch, 1983). In short, the rules are an expression of the knowledge that experts have about a particular type of problem. Because the rules are not mathematically derived functions, a large variety of observations can go into the rules. However, in the best expert systems an attempt is made to distill the expert's knowledge so that only important rules emerge.

Ideally, the user of an expert system is not involved in any aspect of rule development. Instead, the user interacts with the rules via a series of questions. Each question typically requires the user to choose one of several answers. Once the entire series of questions is answered, the rules system is called into action and a decision is made.

As an example, consider a much simplified expert system that determines if an introduced shrub species should be controlled in a nature reserve.

The possible decisions are as follows:

1.0 No management is necessary.
2.0 Control measures are necessary.
2.1 Handcut all individuals at ground level.
2.2 Handcut all individuals and paint with herbicide.

The series of questions asked by the system are as follows:

Q1. Does the aboveground biomass of this plant now exceed $250\,\mathrm{g\,m^{-2}}$?
Q2. Are there rare or endangered species in this management unit?

Q3. Is this shrub known to be allelopathic or does it appear to create bare zones beneath it?
Q4. Does this shrub resprout when cut?

Rules are usually constructed in the form of if-then statements. In this example, only if the user answered yes to Q1, Q2, and Q3 would the system decide that control measures are necessary (decision 2.0). If the user then answered no to Q4, then decision 2.1 would be invoked. A yes to Q4 would invoke decision 2.2.

With a series of four questions this expert system can determine whether control measures should be taken. Granted, this example expert system is oversimplified to the point of limited value, but notice how different types of information are included in the system. Also, take note that the ecological data have been distilled to critical aspects of species performance and interactions with other species. Increasing complexity can be easily brought to expert systems. In the preceding example an obvious addition would be to include statements for managing seedlings of the problem shrub.

A potential drawback to expert systems is availability. To develop an expert system for a specific area, problem, or community type, several types of personnel are needed. First, someone with the ability to build the system – a system planner – is needed. Then the system planner must call together the experts to develop rules. The mere cost of bringing together all these personnel may preclude the development of new expert systems in the future.

Noble (1987) attempted to determine whether expert systems are suitable for use in ecological problems by examining the situations where expert systems perform best (Forsyth, 1984). Specifically, the fact that many ecological problems are diagnostic in nature, they lack a well-developed theory or data base, and often suffer from a lack of qualified personnel, fits expert systems to ecological problems. The only drawback to expert systems is that they may be limited to solving small parts of larger problems (Noble, 1987).

As Starfield and Bleloch (1983) pointed out, expert systems should be designed so that the rules can be modified as more data become available. In this sense, expert systems must be viewed as tools that will improve through time. As with all information systems in resource management, the value or reliability of the output will depend largely on the size and quality of the data base.

9.4 RESOURCE MANAGEMENT ALGORITHMS

As the natural resource management goals set by regulatory agencies and other institutions become increasingly more specific, environmental management becomes exceedingly complex. Systematic procedures are needed that allow resource managers to document, plan, and achieve these goals. Algorithms are well suited to this task and they provide a logical framework for the application of succession knowledge.

An algorithm is a problem-solving approach that involves sequential steps. The algorithm can be designed so that it addresses a relatively well-defined problem or can be sufficiently general so that it can address a wide range of problems. As algorithms get increasingly general and broad, they tend to lose some real-world applicability (see for example Barrett, 1985).

Putman (1988) described the process used in the Mark Twain National Forest (MTNF) in Missouri to meet the guidelines established by the National Forest Management Act. This act requires that each national forest protect species diversity and also determine how management activities are affecting diversity. The approach taken for vertebrate species is that described by Salwasser and Samson (1985) as cumulative effects analysis. The process has four broad steps:

1. A statement of management goals and standards.
2. A representation of habitat factors.
3. A projection of changes in habitat.
4. An estimation of the effects of management.

In the MTNF the first step in this process involved an inventory of animal species and plant communities. Within the plant communities, habitats and habitat elements were identified as well as the numbers and types of animal species supported by each habitat (Table 9.1). Then standards were established. For example, in Table 9.1 a semi-primitive, non-motorized management area in the oak–pine land type is shown regarding the percentage of the management unit now in each habitat and the desired percentages of different habitats that should be there at some point in the future. Desired percentages were set according to land use goals and land capabilities.

For individual management units, habitat factors were identified from a field inventory. These factors included forest type, an assessment of soils and other factors called the ecological land type, productivity, as well as successional stages marked at 10-year intervals (Table 9.2). Such a process in general fulfils the first two steps of the algorithm.

However, habitat factors must be related in some way to vertebrate species diversity. This is accomplished by use of pattern recognition

Table 9.1 Habitats, important elements in these habitats, and standards for an oak–pine land type in the Mark Twain National Forest in Missouri. Existing amounts and desired amounts indicate habitat percentages of the oak–pine land type now in this habitat and the percentages that will encompass this land-type at some point in the future, respectively. Exceptions are permanent water and den trees where existing and desired amounts are expressed as number per unit area

Habitats	Habitat elements	No. of species	Existing amount (%)	Desired amount (%)
1. Forest 0–9 yr old	Forage, open areas	49	8	6–10
2. Old-growth forest >100 yr old	Large diameter stems, mast, dens, logs	270	1	10–15
3. Oak–pine forest >50 yr old	Mast, tree cavities	80	68	40–50
4. Pole and sawtimber	Closed canopies	41	5	35–45
5. Sawtimber with 20–30% ground cover	Moderate understorey	61	72	25–35
6. Oak forest >50 yr old	Dense understorey	74	18	10–15
7. Non-forested habitats	Forage, dense cover	200	<1	1–5
8. Permanent water	Water	—	3 ha^{-1}	3 ha^{-1}
9. Den trees	Tree cavities	155	14 ha^{-1}	>14 ha^{-1}

From Putnam (1988). Reprinted from the Natural Areas Journal with permission of the Natural Areas Association.

Table 9.2 Categorization of forest types and successional stages within each forest type for Kelly Hollow in the Mark Twain National Forest

Forest type	ELT†	Amount (ha) by successional phase‡								
		Grass	Shrub–grass	Shrub–tree	Small trees	Saplings	Poles	Saw-timber	Old growth	Total (ha)
OKHK	Aluv	0	0	0	0	6	47	36	0	89
OKHK	S&W	0	0	0	25	0	44	188	0	258
OKHK	N&E	0	0	0	50	0	23	391	6	470
OKHK	Glade	0	0	0	0	0	2	3	0	5
OPIN	S&W	0	0	0	13	11	96	202	0	321
PINE	S&W	0	0	0	0	0	10	14	0	24
NF	Glade	0	4	0	0	0	0	0	0	4
Total		0	4	0	88	17	222	834	6	1171

* OKHK, oak–hickory; OPIN, oak-pine; NF, non-forested.
† ELT is the Ecological Land Type grouped by slope direction, soil type (alluvial) and other factors.
‡ Each phase represents a period of ca. 10 years.
From Putnam (1988). Reprinted from the Natural Areas Journal with permission of the Natural Areas Association.

computer programs (PATREC). These wildlife habitat evaluation systems predict the suitability of a particular landscape for individual animal species. They do this by comparing the preferred habitat characteristics of a species to the real world mix and juxtaposition of habitats. Output is in the number of animals per unit area that a management unit can support.

With PATREC models the current status of a management unit with regard to overall species diversity or indicator species can be determined. Most importantly, however, PATREC can be linked to a succession model such as DYNAST (Kirkman *et al.*, 1986). DYNAST is a general computer program designed to estimate natural resource production under a variety of management regimes. It includes a routine to predict the mix of various habitats in the future as a result of succession and management activities. By feeding output from DYNAST into PATREC, resource managers in the MTNF can estimate present and future species diversity in a management unit. It is as a succession model that DYNAST achieves its greatest value. The validity of its resource outputs functions, however, have not been well-tested beyond forest communities in the eastern US (Benson and Laudenslayer, 1986).

Both PATREC and DYNAST models allow resource managers to complete the final two steps of the algorithm. Moreover, they can critically examine their management plans so that the habitat mix is brought closer to the established standards.

Cumulative effects analysis is a complex process for inventory and projection of natural resources, but it is a valuable process that recognizes the natural tendency of all plant communities to change, the modification of animal habitat factors associated with this change, as well as the ability of management to deflect or modify succession. Most large nature reserves in the future are likely to use such a process to guide planning and management activities.

SUMMARY

A variety of information systems is available to resource managers so that succession can be considered in decision-making. These systems predict the pathways of succession, predict the effects of management activities, aid personnel in making management decisions and help organize the data collection and presentation procedure to achieve complex management goals.

Succession models simulate plant community changes far into the future. Effects of management activities can also be predicted. Such models are based on mathematical functions or on plant life history characteristics. In general, building a succession model requires extensive knowledge about the system in question. Although many factors controlling plant community change have not been incorporated into models, validation procedures suggest they are relatively good at predicting changes in species importance through time.

An expert system is a tool that can be used by many different types of personnel to make a management decision. It is based on a set of rules developed by a team of experts. The user interacts with the rules by a series of straightforward questions. Expert systems are particularly good for providing resource managers with solutions to small somewhat limited problems.

An algorithm is a systematic procedure that can be used by resource managers to approach complex environmental problems such as assessing the effects of management on species diversity in a large nature reserve. Cumulative effects analysis is one algorithm that includes four steps: stating management goals and standards, representing major habitat factors, projecting changes in habitat, and estimating effects of management on diversity. Completion of this algorithm requires much field-based data. Succession models and wildlife habitat suitability models are used to estimate the outcomes of various management practices.

Afterword

A pervasive thesis of this book is that the rate and trajectory of succession can be manipulated in all plant communities. Few would argue this point. Indeed, it is clear that many management techniques (Chapters 4–7) are now used to manipulate succession to achieve a variety of land use goals (Chapter 1). Still, there is a void that exists between ecologists and resource managers that hinders the development and general acceptance of succession management.

Although ecologists discovered succession and they are constantly elaborating the theoretical aspects of it, foresters, range managers, wildlife biologists, and even landscape architects must manipulate succession with various management techniques as a means to satisfy more applied ends. This demonstrates the utility of succession theory, but also raises the possibility that ecologists will continue to pursue questions about succession without input from those individuals that apply this knowledge in resource management. The result is that resource managers may operate under invalid (outdated) assumptions about succession or they may even ignore it altogether. Moreover, ecologists may ask questions about succession that contribute to the body of knowledge but do not contribute to better resource management.

One solution to this dilemma (Chapter 2) is more co-operation between ecologists and resource managers. Some ecologists might scoff at doing applied research on succession. Instead, they might prefer to work in remote, pristine systems that are not routinely managed. This is truly a luxury as we move into the twenty first century, especially considering the fact that most plant communities on earth have been or are now being manipulated directly or indirectly by people.

The important question is not whether human impact will occur. The answer to this question is painfully obvious as most of us gaze out the windows of our homes or offices. The important question is how can we use our knowledge about succession to attain both biotic preservation and informed resource use. The following suggestions are provided to serve as starting points for greater co-operation between ecologists and resource managers:

1. Conduct more research in anthropic systems. Succession and its manipulation should be studied in agricultural systems, managed forests, rights-of-way, nature reserves, urban parks and even lawns. Managers of these areas should have some input on the types of questions asked by ecologists.
2. Analyse existing standards. Scientific questions about succession should address what standards, if any, are now set for resource use and management. The validity of these standards should be tested within the background of various factors including economic and sociological factors as well as ecological data.
3. Develop novel management strategies and goals. New resource management goals incorporating succession management need to be examined for all public and private lands. At a minimum, natural areas should have policies regarding the manipulation of species diversity in both time and space. Other management goals not typically considered on many public and private lands should be explored: wildlife habitat, scenic vistas, rare species propagation, typal community propagation and restoration, green space, microenvironment amelioration and reduction in fossil fuel use.
4. Implement and monitor novel management strategies on a long-term basis. Management practices tested on small plots must be upscaled and implemented over large areas of land. Long-term monitoring programmes must then be established to track the effects of these strategies.

Obviously, succession will not cease if the previous suggestions are ignored. Instead, succession will continue to function as a repair process following human disturbance just as it has done for the last 5000 years. In many situations this may be adequate resource management. In the majority of situations, however, we need to refine and augment this repair process to better preserve and use our dwindling natural resources.

References

Abrahamson, W.G. (1979) Patterns of resource allocation in wildflower populations of fields and woods. *American Journal of Botany*, **66**, 71–80.

Abrahamson, W.G. and Gadgil, M.D. (1973) Growth form and reproductive effort in goldenrods (*Solidago*, Compositae). *American Naturalist*, **107**, 651–61.

Adams, J.A. and Endo, A.S. (1980) Controlled experiments on soil compaction produced by off-road vehicles in the Mojave Desert, in *The Effects of Disturbance on Desert Soils, Vegetation and Community Processes with Emphasis on Off Road Vehicles: A Critical Review* (ed. P.G. Rowlands), US Department of the Interior Bureau of Land Management, Desert Plan Staff Special Publication, Riverside, California, pp. 121–34.

Altieri, M.A. (1981) Effect of time of disturbance on the dynamics of weed communities in north Florida. *Geobios*, **8**, 145–51.

Anderson, K.L., Smith, E.F. and Owensby, C.E. (1970) Burning bluestem range. *Journal of Range Management*, **23**, 81–92.

Archer, S., Scifres, C., Bassham, C.R. and Maggio, R. (1988) Autogenic succession in a subtropical savanna: conversion of grassland to thorn woodland. *Ecological Monographs*, **58**, 111–27.

Ashby, W.C., Noland, M.R. and Kost, D.A. (1988) *Establishment of Trees in Herbaceous Cover on Graded Lenzburg Minesoil*, US Bureau of Mines Information Circular 9184. pp. 48–53.

Austin, M.P. (1981) Permanent quadrats: an interface for theory and practice. *Vegetatio*, **46**, 1–10.

Bailey, J.A. (1984) *Principles of Wildlife Management*. Wiley, New York, 373 pp.

Bakelaar, R.G. and Odum, E.P. (1978) Community and population level responses to fertilization in an old-field system. *Ecology*, **59**, 660–5.

Bakker, J.P., Brouwer, C., Van Den Hof, L. and Jansen, A. (1987) Vegetational succession, management and hydrology in a brookland (The Netherlands). *Acta Botanica Neerlandica*, **36**, 39–58.

Bakker, J.P., De Bie, S., Dallinga, J.H., Tjaden, P. and DeVries, Y. (1983) Sheep-grazing as a management tool for heathland conservation and regeneration in The Netherlands. *Journal of Applied Ecology*, **20**, 541–60.

Bakker, J.P. and Ruyter, J.C. (1981) Effects of five years of grazing on salt-marsh vegetation. *Vegetatio*, **44**, 81–100.

Barbour, M.G., Burk, J.H. and Pitts, W.D. (1987) *Terrestrial Plant Ecology*. Benjamin Cummings Publishing Company, Menlo Park.

Barrett, G.W. (1985) A problem-solving approach to resource management. *Bioscience*, **35**, 423–7.

Baskin, J.M. and Baskin, C.M. (1979) Studies on the autecology and population biology of the weedy monocarpic perennial, *Pastinaca sativa*. *Journal of Ecology*, **67**, 601–10.

Bayfield, N.G., Urquhart, U.H. and Rothery, P. (1984) Colonization of bulldozed track verges in the Cairngorm Mountains, Scotland. *Journal of Applied Ecology*, **21**, 343–54.

Beckwith, S.L. (1954) Ecological succession on abandoned farm lands and its relationship to wildlife management. *Ecological Monographs*, **24**, 349–76.

Beeftink, W.G. (1979) The structure of salt marsh communities in relation to environmental disturbances, in *Ecological Processes in Coastal Environments* (eds R.L. Jeffries and A.J. Davy), Blackwell, London, pp. 77–93.

Beetle, A.A. (1974) The zootic disclimax concept. *Journal of Range Management*, **27**, 30–2.

Belsky, A.J. (1985) Long-term vegetation monitoring in the Serengeti National Park, Tanzania. *Journal of Applied Ecology*, **22**, 449–60.

Benson, G.L. and Laudenslayer, W.F. Jr (1986) DYNAST: Simulating wildlife responses to forest management strategies, in *Wildlife 2000, Modeling Habitat Relationships of Terrestrial Vertebrates* (eds J. Verner, M.L. Morrison and J. Ralph). University of Wisconsin Press, Madison, Wisconsin, pp. 351–5.

Berglund, B.E. (1969) Vegetation and human influence in south Scandinavia during prehistoric time. *Oikos* supplement, **12**, 9–28.

Biondini, M.E., Bonham, C.D. and Redente, E.F. (1984/1985) Relationships between induced successional patterns and soil biological activity of reclaimed areas. *Reclamation and Revegetation Research*, **3**, 323–42.

Biondini, M.E. and Redente, E.F. (1986) Interactive effect of stimulus and stress on plant community diversity in reclaimed lands. *Reclamation and Revegetation Research*, **4**, 211–22.

Blair, R.M. and Burnett, L.E. (1977) *Deer Habitat Potential of Pine Hardwood Forests in Louisiana*. US Department of Agriculture Forest Service Southern Forest Experiment Station Research Paper SO-136.

Bloomfield, H.E., Handley, J.F. and Bradshaw, A.D. (1982) Nutrient deficiencies and the aftercare of reclaimed derelict land. *Journal of Applied Ecology*, **19**, 151–8.

Bobbink, R. and Willems, J.H. (1987) Increasing dominance of *Brachypodium pinnatum* (L.) Beauv. in chalk grasslands: a threat to a species-rich ecosystem. *Biological Conservation*, **40**, 301–14.

Borman, F.H. and Likens, G.E. (1979) *Pattern and Process in a Forested Ecosystem*. Springer-Verlag, New York, 253 pp.

Boyce, S.G. (1977) *Management of Eastern Hardwood Forests for Multiple Benefits* (DYNAST-MB). United States Department of Agriculture Southeast Forest Experiment Station Research Paper SO-168, 116 pp.

Bradshaw, A.D. and Chadwick, M.J. (1980) *The Ecology and Reclamation of Derelict and Degraded Land*. University of California Press, Berkeley.

Bragg, T.B. and Hulbert, L.C. (1976) Woody plant invasion of unburned Kansas bluestem prairie. *Journal of Range Management*, **29**, 19–24.

Bramble, W.C. and Byrnes, W.R. (1976) Development of a stable, low plant cover on a utility right-of-way. *Proceedings of the First Symposium on Environmental Concerns in Rights-of-way Management*, Mississippi State University, pp. 168–76.

Bramble, W.C. and Byrnes, W.R. (1983) Thirty years of research on development of plant cover on electric transmission right-of-way. *Journal of Arboriculture*, **9**, 67–74.

Bratton, S.P. (1975) The effect of the European wild boar, *Sus scrofa*, on gray beech forest in the Great Smoky Mountains. *Ecology*, **56**, 1356–66.

Bravard, J.P., Amoros, C. and Pautou, G. (1986) Impact of civil engineering works on the successions in a fluvial system. *Oikos*, **47**, 92–111.

Bridgewater, P.B. and Backshall, D.J. (1981) Dynamics of some western Australia ligneous formations with special reference to the invasion of exotic species. *Vegetatio*, **46**, 141–8.

Brockie, R.E., Loope, L.L., Usher, M.B. and Hamann, O. (1988) Biological invasions of island nature reserves. *Biological Conservation*, **44**, 9–36.

Bruns, D. (1986) Does 'reclamation by natural colonization' produce desirable environments? *Proceedings of the 1986 Meeting of the American Society of Surface Mining and Reclamation*, Jackson, Mississippi, pp. 45–50.

Bunting, S.C. and Wright, H.A. (1974) Ignition capabilities on nonflaming firebrands. *Journal of Forestry*, **72**, 646–9.

Bury, R.B. and Luckenbach, R.A. (1983) Vehicular recreation in arid land dunes: biotic responses and management alternatives, in *Environmental Effects of Off-Road Vehicles* (eds R.H. Webb and H.G. Wilshire), Springer-Verlag, New York, pp. 207–21.

Byrnes, W.R. (1961) Vegetation management on utility rights-of-way. *Recent Advances in Botany*, 12, 1426–30.

Cable, D.R. (1972) Fire effects in southwestern semidesert grass–shrub communities. *Proceedings of the Tall Timbers Fire Ecology Conference*, 12, 109–27.

Canham, C.D. and Marks, P.L. (1985) The responses of woody plants to disturbance: patterns of establishment and growth, in *Ecology of Natural Disturbance and Patch Dynamics*, (eds S.T.A. Pickett and P.S. White), Academic Press, London, pp. 197–216.

Cantrell, R.L., Flinchman, D.M. and Neary, D.G. (1986) Individual tree treatments using herbicides for control of turkey oak. *Southern Journal of Applied Forestry*, 10, 131–3.

Carey, A.B. (1985) The scientific basis for spotted owl management, in *Ecology and Management of the Spotted Owl in the Pacific Northwest* (tech eds R.J. Gutierrez and A.B. Carey). United States Forest Service General Technical Report PNW-185, pp. 100–10.

Carrier, W.D. (1985) Planning and managing for the spotted owl in the national forests, Pacific Southwest Region, in *Ecology and Management of the Spotted Owl in the Pacific Northwest* (tech eds R.J. Gutierrez and A.B. Carey). United States Forest Service General Technical Report PNW-185, pp. 2–4.

Carson, W.P. and Barrett, G.W. (1988) Succession in old-field plant communities: effects of contrasting types of nutrient enrichment. *Ecology*, 69, 984–94.

Chapin, F.S. III (1983) Patterns of nutrient absorption and use by plants from natural and man-modified environments, in *Disturbance and Ecosystems* (eds H.A. Mooney and M. Godron). Springer-Verlag, Berlin, pp. 175–87.

Chapman, J., Delcourt, P.A., Criddlebaugh, P.A., Shea, A.B. and Delcourt, H.R. (1982) Man-land interaction: 10,000 years of American indian impact on native ecosystems in the lower Little Tennessee River Valley, eastern Tennessee. *Southeastern Archaeology*, 1, 115–21.

Cheke, A.S., Nanakorn, W. and Yankoses, C. (1979) Dormancy and dispersal of seeds of secondary forest species under the canopy of a primary tropical rain forest in Northern Thailand. *Biotropica*, 11, 88–95.

Christensen, N.L. (1985) Shrubland fire regimes and their evolutionary consequences, in *The Ecology of Natural Disturbance and Patch Dynamics* (eds S.T.A. Pickett and P.S. White). Academic Press, London, pp. 85–100.

Clary, W.P. (1974) Response of herbaceous vegetation to felling of alligator juniper. *Journal of Range Management*, 27, 387–9.

Clements, A.M. (1983) Suburban development and resultant changes in the vegetation of the bushland of the northern Sydney region. *Australian Journal of Ecology*, 8, 307–19.

Clements, F.E. (1916) *Plant succession: an analysis of the development of vegetation.* Carnegie Institute of Washington Publication, 242.

Connell, J.H. and Slatyer, R.O. (1977) Mechanisms of succession in natural communities and their role in community stability and organization. *American Naturalist*, 111, 1119–44.

Conner, W.H., Gosselink, J.G. and Parrondo, R.T. (1981) Comparison of the vegetation of three Louisiana swamp sites with different flooding regimes. *American Journal of Botany*, 68, 320–31.

Cook, R.E. (1985) Growth and development in clonal plant populations, in *Population Biology and Evolution of Clonal Organisms* (eds J.B.C. Jackson, L.W. Buss and R.E. Cook), Yale University Press, New Haven, Connecticut, pp. 259–96.

Cooke, G.D. Welch, E.B., Peterson, S.A. and Newroth, P.R. (1986) *Lake and Reservoir Restoration.* Butterworth Publishers, Boston.

Cowling, R.M., Lamone, B.B. and Pierce, S.M. (1987) Seed bank dynamics of four co-occurring *Banksia* species. *Journal of Ecology*, 75, 289–302.

Crisp, M.D. (1978) Demography and survival under grazing of three Australian semi-desert shrubs. *Oikos*, 30, 520–8.

Davis, B.N.K., Lakhani, K.H. Brown, M.C. and Park, D.G. (1985) Early seral communities in a limestone quarry: an experimental study of treatment effects on cover and richness of vegetation. *Journal of Applied Ecology*, 22, 473–90.

Debussche, M., Escarre, J. and Lepart, J. (1982) Ornithochory and plant succession in mediterranean orchards. *Vegetatio*, 48, 255–66.

Delcourt, H.R. (1987) The impact of prehistoric agriculture and land occupation on natural vegetation. *Trends in Ecology and Evolution*, 2, 39–44.

Depuit, E.J. and Coenenberg, J.G. (1979) *Responses of revegetated coal strip mine spoils to variable fertilization rates, longevity of fertilization program and season of seeding.* Montana Agricultural Experiment Station Research Report No. 150.

Depuit, E.J., Coenenberg, J.G. and Skilbred, C.L. (1980) *Establishment of diverse native plant communities on coal surface-mined lands in Montana as influenced by seeding method, mixture and rate.* Montana Agricultural Experiment Station Research Report No. 163.

Diamond, J.M. and May, R.M. (1976) Island biogeography and the design of nature reserves, in *Theoretical Ecology* (ed. R.M. May), Saunders, Philadelphia, pp. 163–86.

Dickerman, J.A. and Wetzel, R.G. (1985) Clonal growth in *Typha latifolia*: population dynamics and demography of the ramets. *Journal of Ecology*, 73, 535–52.

Dills, G.G. (1970) Effects of prescribed burning on deer browse. *Journal of Wildlife Management*, 34, 540–5.

Dix, R.L. (1957) Sugar maple in forest succession at Washington, DC. *Ecology*, 38, 663–5.

Doerr, T.B. and Redente, E.F. (1983) Seeded plant community changes on intensively disturbed soils as affected by cultural practices. *Reclamation and Revegetation Research*, 2, 13–24.

Doerr, T.B., Redente, E.F. and Reeves, F.B. (1984) Effects of soil disturbance on plant succession and levels of mycorrhizal fungi in a sagebrush-grassland community. *Journal of Range Management*, 37, 135–9.

Dolan, R., Hayden, B.P. and Soucie, G. (1978) Environmental dynamics and resource management in the U.S. national parks. *Environmental Management*, 2, 249–58.

Donelan, M. and Thompson, K. (1980) Distribution of buried viable seeds along a successional series. *Biological Conservation*, 17, 297–311.

Dreyer, G.D. and Niering, W.A. (1986) Evaluation of two herbicide techniques on electric transmission rights-of-way: development of relatively stable shrublands. *Environmental Management*, 10, 113–18.

Drury, W.H. and Nisbet, I.C.T. (1973) Succession. *Journal of the Arnold Arboretum*, 54, 331–68.

Dunn, C.P. and Sharitz, R.R. (1987) Revegetation of a *Taxodium–Nyssa* forested wetland following complete vegetation destruction. *Vegetatio*, 72, 151–7.

Dyer, K.L., Curtis, W.R. and Crews, J.T. (1984) *Responses of vegetation to various mulches used in surface mine reclamation in Alabama and Kentucky – 7-year case history.* United States Department of Agriculture Northeast Forest Experiment Station General Technical Report NE-93.

Edminster, F.C. (1954) *American Game Birds of Field and Forest*, Castle Books, New York, 490 pp.

Egler, F.E. (1949) *Right of Way Maintenance by Plant-community Management*, Aton Forest, Norfolk, Connecticut.

Egler, F.E. (1954) Vegetation science concepts. I. Initial floristic composition – a factor in old-field vegetation development. *Vegetatio*, **4**, 412–17.

Egler, F.E. (1975) *The Plight of the Right of Way Domain – Part I. Victim of Vandalism*. Futura Press, Mt Kisco, New York.

Ehrenfeld, J.G. (1983) The effects of changes in land-use on swamps of the New Jersey Pine Barrens. *Biological Conservation*, **25**, 353–75.

Ehrenfeld, J.G. and Gulick, M. (1981) Structure and dynamics of hardwood swamps in the New Jersey Pine Barrens: contrasting patterns in trees and shrubs. *American Journal of Botany*, **68**, 471–81.

Ellison, L. (1960) Influence of grazing on plant succession in rangelands. *Botanical Review*, **26**, 1–78.

Engle, D.M., Stritzke, J.F. and Claypool, P.L. (1987) Herbage standing crop around eastern redcedar trees. *Journal of Range Management*, **40**, 237–9.

Farmer, R.E. Jr, Cunningham, M. and Barnhill, M.A. (1982) First-year development of plant communities originating from forest topsoils placed on southern Appalachian minesoils. *Journal of Applied Ecology*, **19**, 283–94.

Fenner, M. (1987) Seed characteristics in relation to succession, in *Colonization, Succession and Stability* (eds A.J. Gray, M.J. Crawley and P.J. Edwards), Blackwell Scientific Publications, Oxford, pp. 103–14..

Fonda, R.W. (1974) Forest succession in relation to river terrace development in Olympic National Park, Washington. *Ecology*, **55**, 927–42.

Forman, R.T.T., Galli, A.E. and Leck, C.R. (1976) Forest size and avian diversity in New Jersey woodlots with some land use implications. *Oecologia (Berlin)*, **26**, 1–8.

Forman, R.T.T. and Godron, M. (1986) *Landscape Ecology*. Wiley, New York, 619 pp.

Forsyth, R. (1984) The architecture of expert systems, in *Expert Systems: Principles and Case Studies* (ed. R. Forsyth), Chapman and Hall, London, pp. 9–17.

Fox, M.D. and Fox, B.J. (1986) The susceptibility of natural communities to invasion, in *Ecology of Biological Invasions* (eds R.H. Groves and J.J. Burdon), Cambridge University Press, Cambridge, pp. 57–66.

Franz, E.H. and Bazzaz, F.A. (1977) Simulation of vegetation response to modified hydrologic regimes: A probabilistic model based on niche differentiation in a flood plain forest. *Ecology*, **58**, 176–83.

Fung, M.Y.P. (1986) Ground cover control with herbicides to enhance tree establishment on oil sands reclamation sites. *Proceedings of the 1986 Meeting of the American Society for Surface Mining and Reclamation*, Jackson, Mississippi, pp. 179–82.

Game, M. and Peterken, G.F. (1984) Nature reserve selection strategies in the Woodlands of Central Lincolnshire, England. *Biological Conservation*, **29**, 157–81.

Garcia-Moya, E. and McKell, C.M. (1970) Contribution of shrubs to the nitrogen economy of a desert-wash plant community. *Ecology*, **51**, 81–8.

Gates, J.E. and Gysel, L.W. (1978) Avian nest dispersion and fledging success in field-forest ecotones. *Ecology*, **59**, 871–83.

Gibson, C.W.D. (1986) Management history in relation to changes in the flora of different habitats on an Oxfordshire estate, England. *Biological Conservation*, **38**, 217–32.

Gibson, C.W.D., Watt, T.A. and Brown, V.K. (1987) The use of sheep grazing to recreate species rich grassland from abandoned arable land. *Biological Conservation*, **42**, 165–83.

Gibson, D.J., Johnson, F.L. and Risser, P.G. (1985) Revegetation of unreclaimed coal strip mines in Oklahoma. II. Plant communities. *Reclamation and Revegetation Research*, **4**, 31–47.

Gilbert, O.L. and Wathern, P. (1981) The creation of flower-rich swards on mineral workings. *Reclamation Review*, **3**, 217–21.

Giles, R.H. Jr and Snyder, N. (1970) Simulation techniques in wildlife management, in *Modelling and Systems Analysis in Range Science* (ed. D.A. Jameson). Proceedings of a Symposium of the American Society of Range Management, 9–12 February 1970, Denver Colorado, pp. 23–49.

Gill, A.M. (1981) Fire adaptive traits of vascular plants, in *Fire Regimes and Ecosystem Properties* (eds H.A. Mooney, T.M. Bonnicksen, N.L. Christensen, J.E. Lotan and W.H. Reiners), US Forest Service General Technical Report GTR-WO-26, pp. 208–30.

Gimingham, C.H. (1970) British heathland ecosystems: the outcome of many years of management by fire. *Proceedings of the Tall Timbers Fire Ecology Conference*, **10**, 293–321.

Green, B.H. (1972) The relevance of seral eutrophication and plant competition to the management of successional communities. *Biological Conservation*, **4**, 378–84.

Grime, J.P., Loach, K. and Peckham, D. (1971) Control of vegetation succession by means of soil fabrics. *Journal of Applied Ecology*, **8**, 257–63.

Gross, H.L. (1972) Crown deterioration and reduced growth associated with excessive seed production by birch. *Canadian Journal of Botany*, **50**, 2431–7.

Gross, K.L. (1981) Predictions of fate from rosette size in four 'biennial' plant species: *Verbascum thapsus*, *Oenothera biennis*, *Daucus carota*, and *Tragopogon dubius*. *Oecologia (Berlin)*, **48**, 209–13.

Gross, K.L. and Werner, P.A. (1982) Colonizing abilities of 'biennial' plant species in relation to ground cover: implications for their distributions in a successional sere. *Ecology*, **63**, 921–31.

Grubb, P.J. (1977) The maintenance of species richness in plant communities: the importance of the regeneration niche. *Biological Reviews*, **52**, 107–45.

Hagar, D.C. (1960) The interrelationships of logging, birds and timber regeneration in the douglas-fir region of northwestern California. *Ecology*, **41**, 116–25.

Hall, I.G. (1957) The ecology of disused pit heaps in England. *Journal of Ecology*, **45**, 691–720.

Hall, J.A. (1980) Direct impacts of off-road vehicles on vegetation, in *The Effects of Disturbance on Desert Soils, Vegetation and Community Processes with Emphasis on Off-Road Vehicles: A Critical Review* (ed. P.G. Rowlands), United States Department of the Interior Bureau of Land Management, Desert Plan Staff Special Publication, Riverside, California, pp. 63–74.

Hall, T.F. and Smith, G.E. (1955) Effects of flooding on woody plants, West Sandy dewatering project, Kentucky Reservoir. *Journal of Forestry*, **53**, 281–5.

Hardt, R.A. and Forman, R.T.T. (1989) Boundary form effects on woody colonization of reclaimed surface mines. *Ecology*, **70**, 1252–60.

Hargis, N.E. and Redente, E.F. (1984) Soil handling for surface mine reclamation. *Journal of Soil and Water Conservation*, **39**, 300–5.

Harms, W.R., Schreuder, H.T., Hook, D.D., Brown, C.L. and Shropshire, F.W. (1980) The effects of flooding on the swamp forest in Lake Ocklawaha, Florida. *Ecology*, **61**, 1412–21.

Harper, J.L. (1977) *Population Biology of Plants*. Academic Press, London.

Harper, J.L. and Ogden, J. (1970) The reproductive strategy of higher plants I. The concept of strategy with special reference to *Senecio vulgaris* L. *Journal of Ecology*, **58**, 681–98.

Harper, J.L. and White, J. (1974) The demography of plants. *Annual Review of Ecology and Systematics*, **5**, 419–63.

Harris, S.W. and Marshall, W.H. (1963) Ecology of water-level manipulations on a northern marsh. *Ecology*, **44**, 331–43.

Harrison, J.S. and Werner, P.A. (1984) Colonization by oak seedlings into a heterogeneous successional habitat. *Canadian Journal of Botany*, **62**, 559–63.

Hartnett, D.C. and Bazzaz, F.A. (1983) Physiological integration among interclonal ramets in *Solidago canadensis*. *Ecology*, **64**, 779–88.

Hartnett, D.C. and Bazzaz, F.A. (1985) The genet and ramet population dynamics of *Solidago canadensis* in an abandoned field. *Journal of Ecology*, **73**, 407–13.

Hatton, T.J. and West, N.E. (1987) Early seral trends in plant community diversity on a recontoured surface mine. *Vegetatio*, **73**, 21–9.

Hedin, R.S. (1988) *Volunteer revegetation processes on acid coal spoils in northeastern Pennsylvania*. United States Bureau of Mines Information Circular 9184, pp. 111–17.

Heil, G.W. and Diemont, W.H. (1983) Raised nutrient levels change heathland into grassland. *Vegetatio*, **53**, 113–20.

Heil, G.W., Werger, M.J.A., De Mol, W., Van Dam, D. and Heijne, B. (1988) Capture of atmospheric ammonium by grassland canopies. *Science*, **239**, 764–5.

Henry, J.D. and Swan, J.M.A. (1974) Reconstructing forest history from live and dead plant material – An approach to the study of forest succession in southwest New Hampshire. *Ecology*, **55**, 772–83.

Hickman, J.C. (1975) Environmental unpredictability and plastic energy allocation strategies in the annual *Polygonum cascadense* (polygonaceae). *Journal of Ecology*, **63**, 689–701.

Hobbs, R.J. (1984) Length of burning rotation and composition in high-level *Calluna–Eriophorum* bog in New England. *Vegetatio*, **57**, 129–36.

Hodgkin, S.E. (1984) Scrub encroachment and its effects on soil fertility on Newborough Warren, Anglesey, Wales. *Biological Conservation*, **29**, 99–119.

Hodgkins, E.J. (1958) Effects of fire on undergrowth vegetation in upland southern pine forests. *Ecology*, **39**, 36–46.

Holbrook, H.L. (1974) A system for wildlife habitat management on southern national forests. *Wildlife Society Bulletin*, **2**, 119–23.

Holt, R.B. (1972) Effects of arrival time on recruitment, mortality, and reproduction in successional plant populations. *Ecology*, **53**, 668–73.

Horsley, S.B. and Bjorkman, J.C. (1983) Herbicide treatment of striped maple and beech in Allegheny hardwood stands. *Forest Science*, **29**, 103–12.

Howard, G.S. and Samuel, M.J. (1979) The value of fresh-stripped topsoil as a source of useful plants for surface mine revegetation. *Journal of Range Management*, **32**, 76–7.

Howard-Williams, C. (1975) Vegetation changes in a shallow African lake: response of the vegetation to a recent dry period. *Hydrobiologia*, **47**, 381–98.

Huffine, W.W. and Elder, W.C. (1960) Effects of fertilization on native grass pastures in Oklahoma. *Journal of Range Management*, **13**, 34–6.

Jackson, S.T., Futyma, R.P. and Wilcox, D.A. (1988) A paleoecological test of a classical hydrosere in the Lake Michigan dunes. *Ecology*, **69**, 928–36.

Jacoby, P.W., Slosser, J.E. and Meadors, C.H. (1983) Vegetational responses following control of sand shinnery oak with tebuthiuron. *Journal of Range Management*, **36**, 510–12.

Janzen, D.H. (1976) Why bamboos wait so long to flower. *Annual Review of Ecology and Systematics*, **7**, 347–91.

Jefferson, R.G. and Usher, M.B. (1987) The seed bank in soils of disused chalk quarries in the Yorkshire Wolds, England: Implications for conservation management. *Biological Conservation*, **42**, 287–302.

Jerling, L. (1983) Composition and viability of the seed bank along a successional gradient on a Baltic sea shore meadow. *Holarctic Ecology*, **6**, 150–6.

Johnston, D.W. and Odum, E.P. (1956) Breeding bird populations in relation to plant succession on the piedmont of Georgia. *Ecology*, **37**, 50–62.

Johnston, V.R. (1947) Breeding birds of the forest edge in Illinois. *Condor*, **49**, 45–53.

Joselyn, G.B., Warnock, J.E. and Etter, S.L. (1968) Manipulation of roadside cover for nesting pheasants – a preliminary report. *Journal of Wildlife Management*, **32**, 217–33.

Kachi, N. and Hirose, T. (1985) Population dynamics of *Oenothera glazioviana* in a sand-dune system with special reference to adaptive significance of size-dependent reproduction. *Journal of Ecology*, **73**, 887–901.

Kadlec, J.A. (1962) Effects of a drawdown on a waterfowl impoundment. *Ecology*, **43**, 267–81.

Kayll, A.J. (1974) Use of fire in land management, in *Fire and Ecosystems* (eds T.T. Kozlowski and C.E. Ahlgren), Academic Press, New York, pp. 483–511.

Keddy, P.A., Spavold, A.J. and Keddy, C.J. (1979) Snowmobile impact on old field and marsh vegetation in Nova Scotia, Canada. *Environmental Management*, 3, 409–15.

Keeley, J.E. (1981) Reproductive cycles and fire regimes, in *Fire Regimes and Ecosystem Properties* (eds H.A. Mooney, T.M. Bonnicksen, N.L. Christensen, J.E. Lotan, and W.A. Reiners), United States Forest Service General Technical Report WO-26, pp. 231–77.

Kirkman, R.L., Eberly, J.A., Porath, W.R. and Titus, R.R. (1986) A process for integrating wildlife needs into forest management planning, in *Wildlife 2000, Modeling Habitat Relationships of Terrestrial Vertebrates* (eds J. Verner and M.L. Morrison), University of Wisconsin Press, Madison, pp. 347–50.

Klemow, K.M. and Raynal, D.J. (1983) Population biology of an annual plant in a temporally variable habitat. *Journal of Ecology*, 71, 691–703.

Klemow, K.M. and Raynal, D.J. (1985) Demography of two facultative biennial plant species in an unproductive habitat. *Journal of Ecology*, 73, 147–67.

Kozlowski, T.T. (1971) *Growth and Development of Trees Volume II. Cambial Growth, Root Growth and Reproductive Growth*, Academic Press, New York.

Krefting, L.W. and Phillips, R.S. (1970) Improving deer habitat in upper Michigan by cutting mixed conifer swamps. *Journal of Forestry*, 68, 701–4.

Kroodsma, R.L. (1984) Ecological factors associated with degree of edge effect in breeding birds. *Journal of Wildlife Management*, 48, 418–25.

Kucera, C.L. (1981) Grasslands and fire, in *Fire Regimes and Ecosystem Properties* (eds H.A. Mooney, T.M. Bonnicksen, N.L. Christensen, J.E. Lotan and W.A. Reiners), United States Forest Service General Technical Report WO-26, pp. 90–111.

Lack, D. and Venables, L.S.V. (1939) The habitat distribution of British woodland birds. *Journal of Animal Ecology*, 8, 39–71.

Lathrop, E.W. (1983) Recovery of perennial vegetation in military manoeuvre areas, in *Environmental Effects of Off-Road Vehicles* (eds R.H. Webb and H.G. Wilshire), Springer-Verlag, New York, pp. 265–77.

Leigh, J.H., Winbush, D.J., Wood, D.H., Holgate, M.D., Slee, A.V., Stanger, M.G. and Forrester, R.I. (1987) Effects of rabbit grazing and fire on a subalpine environment I. Herbaceous and shrubby vegetation. *Australian Journal of Botany*, 35, 433–64.

Levenson, J.B. (1981) Woodlots as biogeographic islands in

southeastern Wisconsin, in *Forest Island Dynamics in Man-Dominated Landscapes* (eds R.L. Burgess and D.M. Sharpe), Springer-Verlag, New York, pp. 13–39.

Lewis, C.E. and Harshbarger, T.J. (1976) Shrub and herbaceous vegetation after 20 years of prescribed burning in the South Carolina coastal plain. *Journal of Range Management*, **29**, 13–18.

Lewis, C.E. and Hart, R.H. (1972) Some herbage responses to fire on pine–wiregrass range. *Journal of Range Management*, **25**, 209–13.

Likens, G.E., Bormann, F.H., Johnson, N.M., Fisher, D.W. and Pierce, R.S. (1970) Effects of forest cutting and herbicide treatment on nutrient budgets in the Hubbard Brook watershed ecosystem. *Ecological Monographs*, **40**, 23–47.

Lindsay, M.M. and Bratton, S.P. (1979) Grassy balds of the Great Smoky Mountains: their history and flora in relation to potential management. *Environmental Management*, **3**, 417–30.

Livingston, R.B. and Allessio, M.L. (1968) Buried viable seed in successional field and forest stands, Harvard Forest, Massachusetts. *Bulletin of the Torrey Botanical Club*, **95**, 58–69.

Lloyd, D.G. and Webb, C.J. (1977) Secondary sex characters in plants. *Botanical Review*, **43**, 177–216.

Lovett-Doust, L. (1981) Population dynamics and local specialization in a clonal perennial (*Ranunculus repens*) I. The dynamics of ramets in contrasting habitats. *Journal of Ecology*, **69**, 743–55.

Lowday, J.E. (1987) The effects of cutting and asulam on numbers of frond buds and biomass of fronds and rhizomes of bracken *Pteridium aquilinum*. *Annals of Applied Biology*, **110**, 175–84.

Lowday, J.E., Marrs, R.H. and Nevison, G.B. (1983) Some of the effects of cutting bracken (*Pteridium aquilinum* (L.) Kuhn) at different times during the year. *Journal of Environmental Management*, **17**, 373–80.

Luke, A.G.R., Harvey, H.J. and Humphries, R.N. (1982) The creation of woody landscapes on roadsides by seeding – A comparison of past approaches in West Germany and the United Kingdom. *Reclamation and Revegetation Research*, **1**, 243–53.

Luken, J.O. (1987a) Interactions between seed production and vegetative growth in staghorn sumac, *Rhus typhina* L. *Bulletin of the Torrey Botanical Club*, **114**, 247–51.

Luken, J.O. (1987b) Potential patch dynamics on a roadside embankment: interactions between crownvetch and Kentucky-31 tall fescue. *Reclamation and Revegetation Research*, **6**, 177–86.

Luken, J.O. (1988) Population structure and biomass allocation of the naturalized shrub *Lonicera maackii* (Rupr.) Maxim. in forest and open habitats. *American Midland Naturalist*, **119**, 258–67.

Luken, J.O. and Thieret, J.W. (1987) Sumac-directed patch succession on Northern Kentucky roadside embankments. *Transactions of the Kentucky Academy Science*, **48**, 51–4.

Lyle, E.S. Jr (1988) *Coal surface mine revegetation in the eastern U.S.*, United States Bureau of Mines Information Circular 9184, p. 39.

Macdonald, I.A.W., Graber, D.M., DeBenedetti, S., Groves, R.H. and Fuentes, E.T. (1988) Introduced species in nature reserves in Mediterranean-type climatic regions of the world. *Biological Conservation*, **44**, 37–66.

Mallik, A.U. and Gimingham, C.H. (1983) Regeneration of heathland plants following burning. *Vegetatio*, **53**, 45–58.

Marcot, B.G. (1986) Use of expert systems in wildlife-habitat modeling, in *Wildlife 2000, Modeling Habitat Relationships of Terrestrial Vertebrates* (eds C.J. Verner, M.L. Morrison and C.J. Ralph), University of Wisconsin Press, Madison, pp. 145–50.

Markgren, G. (1974) The moose in Bennoscandia. *Naturaliste Canadien.*, **101**, 185–94.

Marrs, R.H. (1984) Birch control on lowland heaths: mechanical control and the application of selective herbicides by foliar spray. *Journal of Applied Ecology*, **21**, 703–15.

Marrs, R.H. (1985a) *The use of herbicides for nature conservation.* 1985 British Crop Protection Conference – Weeds, pp. 1007–12.

Marrs, R.H. (1985b) Techniques for reducing soil fertility for nature conservation purposes: A review in relation to research at Roper's Heath, Suffolk, England. *Biological Conservation*, **34**, 307–32.

Marrs, R.H. (1987a) Studies on the conservation of lowland *Calluna* heaths I. Control of birch and bracken and its effect on heath vegetation. *Journal of Applied Ecology*, **24**, 163–75.

Marrs, R.H. (1987b) Studies on the conservation of lowland *Calluna* heaths. II. Regeneration of *Calluna* and its relation to bracken infestation. *Journal of Applied Ecology*, **24**, 177–89.

Marrs, R.H., Bravington, M. and Rawes, M. (1988) Long-term vegetation change in the *Juncus squarrosus* grassland at Moor House, northern England. *Vegetatio*, **76**, 179–87.

Marrs, R.H., Hicks, M.J. and Fuller, R.M. (1986) Losses of lowland heath through succession at four sites in Breckland, East Anglia, England. *Biological Conservation*, **36**, 19–38.

McDonnell, M.J. (1986) Old field vegetation height and the dispersal pattern of bird-disseminated woody plants. *Bulletin of the Torrey Botanical Club*, **113**, 6–11.

McDonnell, M.J. and Stiles, E.W. (1983) The structural complexity of old field vegetation and the recruitment of bird-dispersed plant species. *Oecologia (Berlin)*, **56**, 109–16.

McGee, C.E. (1972) From A to. *Journal of Forestry*, 70, 700–4.

McGinnies, W.J. (1987) Effects of hay and straw mulches on the establishment of seeded grasses and legumes on rangeland and a coal strip mine. *Journal of Range Management*, 40, 119–21.

McIntosh, R.P. (1980) The relationship between succession and the recovery process in ecosystems, in *The Recovery Process in Damaged Ecosystems* (ed. J. Cairns Jr), Ann Arbor Science Publishers, Ann Arbor, pp. 11–62.

Medwecka-Kornas, A. (1977) Ecological problems in the conservation of plant communities, with special reference to central europe. *Environmental Conservation*, 4, 27–33.

Meeks, R.L. (1969) The effects of drawdown date on wetland plant succession. *Journal of Wildlife Management*, 33, 817–21.

Mellinger, M.V. and McNaughton, S.J. (1975) Structure and function of successional vascular plant communities in central New York. *Ecological Monographs*, 45, 161–82.

Miles, J. (1974) Effects of experimental interference with stand structure on establishment of seedlings in Callunetum. *Journal of Ecology*, 62, 675–87.

Miles, J. (1988) Vegetation and soil change in the uplands, in *Ecological Change in the Uplands* (eds M.B. Usher and D.B.A. Thompson), Blackwell Scientific Publications, Oxford, pp. 57–69.

Miller, G.R. and Miles, J. (1970) Regeneration of heather (*Calluna vulgaris* L. Hull) at different ages and seasons in north-east Scotland. *Journal of Applied Ecology*, 7, 51–60.

Miller, G.R., Watson, A. and Jenkins, D. (1970) Responses of red grouse populations to experimental improvement of their food, in *Animal Populations in Relation to Their Food Resources* (ed. A. Watson), Blackwell Scientific Publications, Oxford, pp. 323–35.

Milton, K., Windsor, D.M. and Morrison, D.W. (1982) Fruiting phenologies of two neotropical *Ficus* species. *Ecology*, 63, 752–62.

Mobley, H.E., Jackson, R.S., Balmer, W.E., Ruziska, W.E. and Hough, W.A. (1973) *A Guide for Prescribed Fire in Southern Forests*, US Forest Service Southern Region Rev. 1978.

Moore, W.H., Swinder, B.F. and Terry, W.S. (1982) Vegetative response to clearcutting and chopping in North Florida flatwoods forest. *Journal of Range Management*, 35, 214–18.

Moss, R., Miller, G.R. and Allen, S.E. (1972) Selection of heather by captive red grouse in relation to the age of the plant. *Journal of Applied Ecology*, 9, 771–81.

Mroz, G.D., Frederick, D.J. and Jurgensen, M.F. (1985) Site and fertilizer effects on northern hardwood stump sprouting. *Canadian Journal of Forest Research*, 15, 535–43.

Mueller-Dombois, D. (1981) Vegetation dynamics in a coastal grassland of Hawaii. *Vegetatio*, **46**, 131–40.

Mueller-Dombois, D., Canfield, J.E., Holt, R.A. and Buelow, G.P. (1983) Tree-group death in North America and Hawaiian forests: a pathological problem or a new problem for vegetation ecology? *Phytocoenolgia*, **11**, 117–37.

Muller, R.N. (1982) Vegetation patterns in the mixed mesophytic forest of eastern Kentucky. *Ecology*, **63**, 1901–17.

Naylor, R.E.L. (1985) Establishment and peri-establishment mortality, in *Studies on Plant Demography: A Festschrift for John L. Harper* (ed. J. White), Academic Press, London, pp. 95–107.

Newell, S.J. and Tramer, E.J. (1978) Reproductive strategies in herbaceous plant communities during succession. *Ecology*, **59**, 228–34.

Niemi, G.J. and Hanowski, J.M. (1984) Relationships of breeding birds to habitat characteristics in logged areas. *Journal of Wildlife Management*, **48**, 438–43.

Niering, W.A. (1987) Vegetation dynamics (succession and climax) in relation to plant community management. *Conservation Biology*, **1**, 287–95.

Niering, W.A., Dreyer, G.D., Egler, F.E. and Anderson, J.P. (1986) Stability of a *Viburnum lentago* shrub community after 30 years. *Bulletin of the Torrey Botanical Club*, **113**, 23–7.

Niering, W.A. and Egler, F.E. (1955) A shrub community of *Viburnum lentago*, stable for twenty-five years. *Ecology*, **36**, 356–60.

Niering, W.A. and Goodwin, R.H. (1974) Creation of relatively stable shrublands with herbicides: arresting succession on rights-of-way and pastureland. *Ecology*, **55**, 784–95.

Nilsson, C. (1981) Dynamics of the shore vegetation of a north Swedish hydro-electric reservoir during a 5-year period. *Acta Phytogeographica Suecica*, **69**, 1–94.

Noble, I.R. (1987) The role of expert systems in vegetation science. *Vegetatio*, **69**, 115–21.

Noble, I.R. and Slatyer, R.O. (1977) Post-fire succession of plants in Mediterranean ecosystems, in *Proceedings of the Symposium on the Environmental Consequences of Fire and Fuel Management in Mediterranean Ecosystems*, United States Forest Service General Technical Report WO-3, pp. 27–36.

Noble, I.R. and Slatyer, R.O. (1980) The use of vital attributes to predict successional changes in plant communities subject to recurrent disturbances. *Vegetatio*, **43**, 5–21.

Noble, J.C., Bell, A.D. and Harper, J.L. (1979) The population biology of plants with clonal growth. I. The morphology and structural demography of *Carex arenaria*. *Journal of Ecology*, **67**, 983–1008.

Norris, L.A. (1967) The physiological and biochemical bases of selective herbicide action, *Proceedings of a Symposium: Herbicides and Vegetation Management in Forests, Ranges and Noncrop Lands*, pp. 56–66.

Noss, R.F. and Harris, L.D. (1986) Nodes, networks and MUMs: preserving diversity at all scales. *Environmental Management*, **10**, 299–309.

Oomes, M.J.M. and Mooi, H. (1981) The effect of cutting and fertilizing on the floristic composition and production of an *Arrhenatherion elatioris* grassland. *Vegetatio*, **47**, 233–9.

Oosting, H.J. and Humphreys, M.E. (1940) Buried viable seeds in a successional series of old field and forest soils. *Bulletin of the Torrey Botanical Club*, **67**, 253–73.

Owen, J.S. (1972) Some thoughts on management in national parks. *Biological Conservation*, **4**, 241–6.

Owensby, C.E., Hyde, R.M. and Anderson, K.L. (1970) Effects of clipping and supplemental nitrogen and water on loamy upland bluestem range. *Journal of Range Management*, **23**, 341–6.

Palaniappan, V.M., Marrs, R.W. and Bradshaw, A.D. (1979) The effect of *Lupinus arboreus* on the nitrogen status of china clay wastes. *Journal of Applied Ecology*, **16**, 825–31.

Peek, J.M., Urich, D.L. and Mackie, R.J. (1976) Moose habitat selection and relationships to forest management in northeastern Minnesota. *Wildlife Monographs*, **48**, 1–65.

Peet, R.K. and Christensen, N.L. (1980) Succession: A population process. *Vegetatio*, **43**, 131–40.

Pendleton, R.F. (1983) Some herbicides for watersheds and roadside rights-of-way. *Journal of Arboriculture*, **9**, 263–6.

Persson, H. (1981) The effect of fertilization and irrigation on the vegetation dynamics of a pine–heath ecosystem. *Vegetatio*, **46**, 181–92.

Petranka, J.W. and McPherson, J.K. (1979) The role of *Rhus copallina* in the dynamics of the forest–prairie ecotone in north-central Oklahoma. *Ecology*, **60**, 956–65.

Pickett, S.T.A. (1976) Succession: an evolutionary interpretation. *American Naturalist*, **110**, 107–19.

Pickett, S.T.A. (1982) Population patterns through twenty years of oldfield succession. *Vegetatio*, **49**, 45–59.

Pickett, S.T.A., Collins, S.L. and Armesto, J.J. (1987) Models, mechanisms and pathways of succession. *The Botanical Review*, **53**, 335–71.

Pickett, S.T.A. and Thompson, J.N. (1978) Patch dynamics and the design of nature reserves. *Biological Conservation*, **13**, 27–37.

Pineda, F.D., Casado, M.A., Peco, B., Olmeda, C. and Levassor, C. (1987) Temporal changes in therophytic communities across the boundary of disturbed-intact ecosystems. *Vegetatio*, **71**, 33–9.

Piñero, D., Sarukhán, J. and Alberdi, P. (1982) The cost of reproduction in a tropical palm, *Astrocaryum mexicanum*. *Journal of Ecology*, **70**, 473–81.

Pitelka, L.F. and Ashmun, J.W. (1985) Physiology and integration of ramets in clonal plants, in *Population biology and evolution of clonal organisms* (eds J.B.C. Jackson, L.W. Buss and R.W. Cook), Yale University Press, New Haven, pp. 399–435.

Pound, C.E. and Egler, F.E. (1953) Brush control in southeastern New York: fifteen years of stable treeless communities. *Ecology*, **34**, 63–73.

Powell, T.A., Guarino, L. and Harvey, H.J. (1985) The experimental management of vegetation at Wicken Fen, Cambridgeshire. *Journal of Applied Ecology*, **22**, 217–27.

Prose, D.V., Metzger, S.K. and Wilshire, H.G. (1987) Effects of substrate disturbance on secondary plant succession; Mojave Desert, California. *Journal of Applied Ecology*, **24**, 305–13.

Putnam, C. (1988) The development and application of habitat standards for maintaining vertebrate species diversity on a national forest. *Natural Areas Journal*, **8**, 256–66.

Rafe, R.W., Usher, M.B. and Jefferson, R.G. (1985) Birds on reserves: the influence of area and habitat on species richness. *Journal of Applied Ecology*, **22**, 327–35.

Ranney, J.W., Bruner, M.C. and Levenson, J.B. (1981) The importance of edge in the structure and dynamics of forest islands, in *Forest Island Dynamics in Man-Dominated Landscapes* (eds R.L. Burgess and D.M. Sharpe), Springer-Verlag, New York, pp. 67–95.

Redente, E.F., Doerr, T.B., Grygiel, C.E. and Biondini, M.E. (1984) Vegetation establishment and succession on disturbed soils in northwest Colorado. *Reclamation and Revegetation Research*, **3**, 153–65.

Reekie, E.G. and Bazzaz, F.A. (1987) Reproductive effort in plants 1. Carbon allocation to reproduction. *American Naturalist*, **129**, 876–96.

Rehm, G.W., Moline, W.J. and Schwartz, E.J. (1972) Response of a seeded mixture of warm-season prairie grasses to fertilization. *Journal of Range Management*, **25**, 452–6.

Rice, B. and Westoby, M. (1978) Vegetative responses of some Great Basin shrub communities protected against jackrabbits or domestic stock. *Journal of Range Management*, **31**, 28–34.

Ringe, J.M. and Graves, D.H. (1987) Economic factors affecting mulch choices for revegetating disturbed land. *Reclamation and Revegetation Research*, **6**, 121–8.

Rippel, P., Pieper, R.D. and Lymbery, G.A. (1983) Vegetational evaluation of pinyon–juniper cabling in south-central New Mexico. *Journal of Range Management*, **36**, 13–5.

Roberts, R.D., Marrs, R.H., Skeffington, R.A. and Bradshaw, A.D. (1981) Ecosystem development on naturally colonized china clay wastes. I. Vegetation changes and overall accumulation of organic matter and nutrients. *Journal of Ecology*, **69**, 153–61.

Rorison, I.H. (1971) The use of nutrients in the control of the floristic composition of grassland, in *The Scientific Management of Animal and Plant Communities for Conservation* (eds E. Duffey and A.S. Watt), Blackwell Scientific Publications, Oxford, pp. 65–77.

Rose Innes, R. (1971) Fire in West African vegetation. *Proceedings of the Tall Timbers Fire Ecology Conference*, **11**, 147–73.

Rosenberg, D.B. and Freedman, S.M. (1984) Application of a model of ecological succession to conservation and land-use management. *Environmental Conservation*, **11**, 323–9.

Rosenberg, K.V. and Raphael, M.G. (1986) Effects of forest fragmentation on vertebrates in douglas-fir forests, in *Wildlife 2000: Modeling Habitat Relationships of Terrestrial Vertebrates* (eds J. Verner, M.L. Morrison and C.J. Ralph), University of Wisconsin Press, Madison, pp. 263–72.

Sakai, A.K. and Sulak, J.H. (1985) Four decades of secondary succession in two lowland permanent plots in northern lower Michigan. *American Midland Naturalist*, **113**, 146–57.

Salisbury, E.J. (1942) *The Reproductive Capacity of Plants*, Bell, London.

Salwasser, H. and Samson, F.B. (1985) Cumulative effects analysis: an advance in wildlife planning and management, in *Transactions of the Fiftieth North American Wildlife and Natural Resource Conference*, Wildlife Management Institute, Washington DC, pp. 313–21.

Schaller, F.W. and Sutton, P. (eds) (1978) *Reclamation of Drastically Disturbed Lands*, American Society of Agronomy, Madison, Wisconsin.

Schier, G.A., Sheppard, W.D. and Jones, J.R. (1985) Regeneration, in *Aspen: Ecology and Management in the Western United States* (eds N.V. Debyle and R.R. Winokur), United States Forest Service Rocky Mountain Forest and Range Experiment Station Report GTM RM-119, pp. 197–208.

Schier, G.A. and Smith, A.D. (1979) *Sucker regeneration in Utah aspen clones following clearcutting, partial cutting and girdling.* United States Forest Service Intermountain Forest and Range Experiment Station Research Note INT-253.

Schmutz, E.M., Cable, D.R. and Warwick, J.J. (1959) Effects of shrub removal on the vegetation of a semidesert grass–shrub range. *Journal of Range Management*, **12**, 34–7.

Schott, M.R. and Pieper, R.D. (1987) Succession of pinyon–juniper communities after mechanical disturbance in southcentral New Mexico. *Journal of Range Management*, **40**, 88–94.

Schroeder, P.M., Dolan, R. and Hayden, B.P. (1976) Vegetation changes associated with barrier-dune construction on the Outer Banks of North Carolina. *Environmental Management*, **1**, 105–14.

Schulze, E.D., Kuppers, M. and Matyssek, R. (1986) The roles of carbon balance and branching pattern in the growth of woody species, in *On the Economy of Plant Form and Function* (ed. T.J. Givnish), Cambridge University Press, Cambridge, pp. 585–602.

Schuster, W.S. and Hutnick, R.J. (1987) Community development on 35-year-old planted minespoil banks in Pennsylvania. *Reclamation and Revegetation Research*, **6**, 109–20.

Sheldon, J.D. and Bradshaw, A.D. (1977) The development of a hydraulic seeding technique for unstable sand slopes I. Effects of fertilizers, mulches and stabilizers. *Journal of Applied Ecology*, **14**, 905–18.

Shugart, H.H. (1984) *A Theory of Forest Dynamics: The Ecological Implications of Forest Succession Models*, Springer Verlag, New York.

Shugart, H.H. Jr, Crow, T.R. and Hett, J.M. (1973) Forest succession models: a rationale and methodology for modeling forest succession over large regions. *Forest Science*, **19**, 203–12.

Shugart, H.H. Jr, Hopkins, M.S., Burgess, I.P. and Mortlock, A.T. (1980) The development of a succession model for subtropical rain forest and its applications to assess the effects of timber harvest at Wiangaree State Forest, New South Wales. *Journal of Environmental Management*, **11**, 243–65.

Shugart, H.H. Jr and West, D.C. (1977) Development of an Appalachian deciduous forest succession model and its application to assessment of the impact of the chestnut blight. *Journal of Environmental Management*, **5**, 161–79.

Silvertown, J.W. (1980) The evolutionary ecology of mast seeding in trees. *Biological Journal of the Linnean Society*, **14**, 235–50.

Simberloff, D. (1986) Design of nature reserves, in *Wildlife Conservation Evaluation* (ed. M.B. Usher), Chapman and Hall, London, pp. 315–37.

Slatyer, R.O. (1977) *Dynamic Changes in Terrestrial Ecosystems: Patterns of Change, Techniques for Study and Applications to Management*, Workshop Findings Sponsored by UNESCO-MAB and ICSU-SCOPE, Santa Barbara, California, January, 1976.

Slick, B.M. and Curtis, W.R. (1985) *A Guide for the Use of Organic Materials in Mulches in Reclamation of Coal Minesoils in the Eastern United States*. United States Forest Service Northeastern Experiment Station General Technical Report NE-98.

Smart, N.O.E., Hatton, J.C. and Spence, D.H.N. (1985) The effect of long-term exclusion of large herbivores on vegetation in Murchison Falls National Park, Uganda. *Biological Conservation*, **33**, 229–45.

Smith, D.M. (1986) *The Practice of Silviculture*. John Wiley and Sons, New York.

Smith, E.F. and Owensby, C.E. (1972) Effects of fire on true prairie grassland. *Proceedings of the Tall Timbers Fire Ecology Conference*, **12**, 9–22.

Smith, L.M. and Kadlec, J.A. (1983) Seed banks and their role during drawdown of a North American marsh. *Journal of Applied Ecology*, **20**, 673–84.

Smith, R.S. (1988) Farming and the conservation of traditional meadowland in the Pennine Dales environmentally sensitive area, in *Ecological Change in the Uplands* (eds M.B. Usher and D.B.A. Thompson). Blackwell Scientific Publications, Oxford, pp. 183–99.

Smith, W.H. (1981) *Air Pollution and Forests*. Springer-Verlag, New York, 377 pp.

Sork, V.L. (1983) Mast-fruiting in hickories and availability of nuts. *American Midland Naturalist*, **109**, 81–8.

Soutiere, E.C. (1979) Effects of timber harvesting on marten in Maine. *Journal of Wildlife Management*, **43**, 850–60.

Specht, R.L., Connor, D.J. and Clifford, H.T. (1977) The heath–savannah problem: the effects of fertilizer on sand–heath vegetation of North Stradbroke Island, Queensland. *Australian Journal of Ecology*, **2**, 179–86.

Spence, D.H.N. (1982) The zonation of plants in freshwater lakes. *Advances in Ecological Research*, **12**, 37–125.

Starfield, A.M. and Bleloch, A.L. (1983) Expert systems: an approach to problems in ecological management that are difficult to quantify. *Journal of Environmental Management*, **16**, 261–8.

Sterrett, J.P. and Adams, R.E. (1977) The effect of forest conversion and herbicides on pine (*Pinus* spp.) establishment, soil moisture and understory vegetation. *Weed Science*, **25**, 521–3.

Strelke, W.K. and Dickson, J.G. (1980) Effects of forest clear-cut edge on breeding birds in east Texas. *Journal of Wildlife Management*, **44**, 559–67.

Telfer, E.S. (1974) Logging as a factor in wildlife ecology in the boreal forest. *Forestry Chronicle*, **50**, 186–90.

Thibodeau, F.R. and Nickerson, N.H. (1985) Changes in a wetland plant association induced by impoundment and draining. *Biological Conservation*, **33**, 269–79.

Thilenius, J.F. and Brown, G.R. (1974) Long-term effects of chemical control of big sagebrush. *Journal of Range Management*, **27**, 223–4.

Thompson, F.R. and Capen, D.E. (1988) Avian assemblages in seral stages of a Vermont forest. *Journal of Wildlife Management*, **52**, 771–7.

Thompson, J.N. and Willson, M.F. (1979) Evolution of temperate fruit/bird interactions: Phenological strategies. *Evolution*, **33**, 973–98.

Thompson, R.L., Vogel, W.G. and Taylor, D.D. (1984) Vegetation and flora of a coal surface-mined area in Laurel County, Kentucky. *Castanea*, **49**, 111–26.

Thompson, R.L., Vogel, W.G., Wade, G.L. and Refaili, B.L. (1986) Development of natural and planted vegetation on surface mines in southeastern Kentucky. *Proceedings of the National Meeting of the American Society of Surface Mining and Reclamation*, Jackson, Mississippi, pp. 145–54.

Tiedemann, A.R. and Klemmedson, J.O. (1977) Effect of mesquite trees on vegetation and soils in the desert grassland. *Journal of Range Management*, **30**, 361–7.

Tiffney, B.H. and Niklas, K.J. (1985) Clonal growth in land plants. A paleobotanical perspective, in *Population Biology and Evolution of Clonal Organisms* (eds J.B.C. Jackson, L.W. Buss and R.E. Cook), Yale University Press, New Haven, pp. 35–66.

Tilman, G.D. (1984) Plant dominance along an experimental nutrient gradient. *Ecology*, **65**, 1445–53.

Tracey, W.H. and Glossop, B.L. (1980) Assessment of topsoil handling techniques for rehabilitation of sites mined for bauxite within the jarrah forest of western Australia. *Journal of Applied Ecology*, **17**, 195–201.

Uhl, C. (1987) Factors controlling succession following slash-and-burn agriculture in Amazonia. *Journal of Ecology*, **75**, 377–407.

Uhl, C., Jordan, C., Clark, K., Clark, H. and Herrera, R. (1982) Ecosystem recovery in Amazon caatinga forest after cutting, cutting and burning, and bulldozer clearing treatments. *Oikos*, **38**, 313–20.

Usher, M.B. (1988) Biological invasions of nature reserves: a search for generalizations. *Biological Conservation*, **44**, 119–35.

van der Maarel, E. (1978) Experimental succession research in a coastal dune grassland, a preliminary report. *Vegetatio*, **38**, 21–8.

van der Meulen, F. (1982) Vegetation changes and water catchment in a dutch coastal dune area. *Biological Conservation*, **24**, 305–16.

van der Pijl, L. (1972) *Principles of Dispersal in Higher Plants*, Springer Verlag, Berlin.

van der Valk, A.G. (1987) Vegetation dynamics of freshwater wetlands: A selective review of the literature. *Archiv. für Hydrobiologie*, **27**, 27–39.

van der Valk, A.G. and Davis, C.B. (1978) The role of seed banks in the vegetation dynamics of prairie glacial marshes. *Ecology*, **59**, 322–35.

Van Wyk, P. (1971) Veld burnings in the Kruger National Park. An interim report of some aspects of research. *Proceedings of the Tall Timbers Fire Ecology Conference*, **11**, 9–31.

Vance, D.T. (1976) Changes in land use and wildlife populations in southwestern Illinois. *Wildlife Society Bulletin*, **4**, 11–15.

Verme, L.J. (1965) Swamp conifer deer yards in northern Michigan: their ecology and management. *Journal of Forestry*, **63**, 523–9.

Vestergaard, P. (1985) Effects of mowing on the composition of Baltic salt-meadow communities. *Vegetatio*, **62**, 383–90.

Vinther, E. (1983) Invasion of *Alnus glutinosa* (L.) Gaertn. in a former grazed meadow in relation to different grazing intensities. *Biological Conservation*, **25**, 75–89.

Vitousek, P.M. and Reiners, W.A. (1975) Ecosystem succession and nutrient retention: A hypothesis. *Bioscience*, **25**, 376–81.

Vogel, W.G. and Berg, W.A. (1973) Fertilizer and herbaceous cover influence establishment of direct-seeded black locust on coal-mine spoils, in *Ecology and Reclamation of Devastated Land* (eds R.J. Hutnik and G. Davis), pp. 189–97. Gordon and Breach, New York.

Vogl, R.J. (1972) Fire in southeastern grasslands. *Proceedings of the Tall Timbers Fire Ecology Conference*, **12**, 175–93.

Vollmer, A.T., Maza, B.G., Medica, P.A., Turner, F.B. and Bamberg, S.A. (1976) The impact of off-road vehicles on a desert ecosystem. *Environmental Management*, **1**, 115–29.

Wade, G.L. (1989) Grass competition and establishment of native species from forest soil seed banks. *Landscape and Urban Planning*, **17**, 135–49.

Walker, L.R. and Chapin, F.S. III (1987) Interactions among processes controlling successional change. *Oikos*, **50**, 131–5.

Walker, L.R., Zasada, J.C. and Chapin, F.S. III (1986) The role of life history processes in primary succession on an Alaskan floodplain. *Ecology*, **67**, 1243–53.

Wambolt, C.L. and Payne, G.F. (1986) An 18-year comparison of control methods for Wyoming big sagebrush in southwestern Montana. *Journal of Range Management*, **39**, 314–19.

Warner, R.E. and Joselyn, G.B. (1986) Responses of Illinois ring-necked pheasant populations to block roadside management. *Journal of Wildlife Management*, **50**, 525–32.

Watt, A.S. (1957) The effect of excluding rabbits from grassland B (Mesobrometum) in Breckland. *Journal of Ecology*, **45**, 861–78.

Watt, A.S. (1960a) Population change in acidophilous grass–heath in Breckland 1936–1957. *Journal of Ecology*, **48**, 605–29.

Watt, A.S. (1960b) The effect of excluding rabbits from acidophilous grassland in Breckland, 1936–57. *Journal of Ecology*, **48**, 601–4.

Watt, A.S. (1962) The effect of excluding rabbits from grassland A (Zerobrometum) in Breckland, 1936–60. *Journal of Ecology*, **50**, 181–98.

Webb, R.H., Wilshire, H.G. and Henry, H.A. (1983) Natural recovery of soils and vegetation following human disturbance, in *Environmental Effects of Off-Road Vehicles* (eds R.W. Webb and H.G. Wilshire), Springer-Verlag, New York, pp. 279–302.

Webb, W.L., Behrend, D.G. and Saisorn, B. (1977) Effect of logging on songbird populations in a northern hardwood forest. *Wildlife Monographs*, **55**, 1–35.

Weber, J.B., Best, J.A. and Witt, W.W. (1974) Herbicide residues and weed species shifts on modified-soil field plots. *Weed Science*, **22**, 427–33.

Wegner, J.F. and Merriam, G. (1979) Movements of birds and small mammals between a wood and adjoining farmland habitats. *Journal of Applied Ecology*, **16**, 349–58.

Welch, D. (1984) Studies in the grazing of heather moorland in north-east Scotland. III. Floristics. *Journal of Applied Ecology*, **21**, 209–25.

Welch, D. (1985) Studies in the grazing of heather moorland in north-east Scotland IV. Seed dispersal and plant establishment in dung. *Journal of Applied Ecology*, **22**, 461–72.

Wells, T.C.E. (1971) A comparison of the effects of sheep grazing and mechanical cutting on the structure and botanical composition of chalk grassland, in *The Scientific Management of Animal and Plant Communities* (eds E. Duffey and A.S. Watt), Blackwell Scientific Publications, Oxford, pp. 497–515.

Werner, P.A. (1975) The effects of plant litter on germination in teasel, *Dipsacus sylvestris* Huds. *American Midland Naturalist*, **94**, 470–6.

Werner, P.A. and Harbeck, A.L. (1982) The pattern of tree seedling establishment relative to staghorn sumac cover in Michigan old fields. *American Midland Naturalist*, **108**, 124–32.

West, D.C., McLaughlin, S.B. and Shugart, H.H. (1980) Simulated forest response to chronic air pollution stress. *Journal of Environmental Quality*, **9**, 43–9.

Westhoff, V. (1971) The dynamic structure of plant communities in relation to the objectives of conservation, in *The Scientific Management of Animal and Plant Communities* (eds E. Duffey and A.S. Watt). Blackwell Scientific Publications, Oxford, pp. 3–14.

Whipple, S.A. and Dix, R.L. (1979) Age structure and successional dynamics of a Colorado subalpine forest. *American Midland Naturalist*, **101**, 142–58.

White, P.S. (1979) Pattern, process and natural disturbance in vegetation. *Botanical Review*, **45**, 229–99.

White, P.S. and Bratton, S.P. (1980) After preservation: philosophical and practical problems of change. *Biological Conservation*, **18**, 241–55.

White, P.S. and Pickett, S.T.A. (1985) Natural disturbance and patch dynamics, in *The Ecology of Natural Disturbance and Patch Dynamics* (eds S.T.A. Pickett and P.S. White), Academic Press, Orlando, pp. 3–13.

Whitney, G.G. and Runkle, J.R. (1981) Edge versus age effects in the development of a beech–maple forest. *Oikos*, **37**, 377–81.

Whittaker, R.H. (1975) *Communities and Ecosystems*, Macmillan Publishing, New York.

Williams, R.F. (1955) Redistribution of mineral elements during development. *Annual Review of Plant Physiology*, **6**, 25–42.

Williams, R.J. and Ashton, D.H. (1987) Effects of disturbance and grazing by cattle on the dynamics of heathland and grassland communities on the Bogong High Plains, Victoria. *Australian Journal of Botany*, **35**, 413–31.

Willis, A.J. (1963) Braunton Burrows: The effects on the vegetation of the addition of mineral nutrients to the dune soils. *Journal of Ecology*, 51, 353–74.

Willson, M.F. (1986) Avian frugivory and seed dispersal in eastern North America. *Current Ornithology*, 3, 223–79.

Winbush, D.J. and Costin, A.B. (1979) Trends in vegetation at Kosciusko. I. Grazing trials in the subalpine zone, 1957–1971. *Australian Journal of Botany*, 27, 741–87.

Wittwer, R.F., Graves, D.H. and Carpenter, S.B. (1979) Establishing oaks and virginia pine on Appalachian surface mine spoils by direct seeding. *Reclamation Review*, 2, 63–6.

Wood, M.K. and Blackburn, W.H. (1984) Vegetation and soil responses to cattle grazing systems in the Texas rolling plains. *Journal of Range Management*, 37, 303–8.

Wright, D.L., Perry, H.D. and Blaser, R.E. (1978) Persistent low maintenance vegetation for erosion control and aesthetics in highway corridors, in *Reclamation of Drastically Disturbed Lands* (eds F.W. Schaller and P. Sutton), American Society of Agronomy, Crop Science Society of America, and Soil Science Society of America, Madison, Wisconsin, pp. 553–83.

Wright, H.A. and Bailey, A.W. (1982) *Fire Ecology United States and Canada*, Wiley, New York.

Yeiser, J.L. (1986) Tree injection for early pine seedling release in the Ozark Mountains of Arkansas. *Southern Journal of Applied Forestry*, 10, 249–51.

Young, K.R., Ewel, J.J. and Brown, B.J. (1987) Seed dynamics during forest succession in Costa Rica. *Vegetatio*, 71, 157–73.

Zasada, J. (1986) Natural regeneration of trees and tall shrubs on forest sites in interior Alaska, in *Forest Ecosystems in the Alaskan Taiga* (eds K. VanCleve, F.S. Chapin III, P.W. Flanagan, L.A. Viereck and C.T. Dyrness), Springer Verlag, New York, pp. 44–73.

Zcdaker, S.M., Lewis, J.B., Smith, D.W. and Kreh, R.E. (1987) Impact of season of harvest and site quality on cut-stump treatments of piedmont hardwoods. *Southern Journal of Applied Forestry*, 11, 46–9.

Index

Acacia sieberiana, 175
Acalypha bipartita, 175
Acer rubrum, 124
Achyranthes aspera, 175
Age structure
   and population trends, 38–9
Agropyron cristatum, 109
Agropyron spicatum, 83
Agrostis stolonifera, 110
Air pollution, 197–8
Alces alces, 161–2
Algorithms, 211–15, 216
Allocation
   and seed production, 47–8
   and vegetative reproduction, 54
Alopecurus pratensis, 110
Alnus rubra, 29
Alnus tenuifolia, 46
Ambrosia artemisiifolia, 26
Ambrosia dumosa, 196
Andropogon gerardii, 88, 115
Anecdotal information, 30, 210
Animal migration, 179, 190, 191
Animal succession, 151
Anthropic communities, 179–80
Applied research, 19, 20, 217
Arrested succession, 77–80, 132
Artemisia tridentata, 83
Artificial flooding, 121–4
Artificial perches
   and tree invasion, 177, 178
Asexual reproduction, 37, 54, 57
Aster pilosus, 105
Astragalus cicer, 109
Atriplex canescens, 138

Barrier islands, 197
Betula pendula, 64, 91
Bossiaea foliosa, 169
Brachypodium pinnatum, 198
Bromus inermis 109
Bromus japonicus, 174
Browse availability, 159, 161, 162, 178
Buchloe dactyloides, 174

Cabling, 97–100, 101
Calluna vulgaris, 64–5, 83, 89–91, 108, 117, 159, 167
Cambretum binderianum, 176
Capra hircus, 193
Carex nigra, 167
Castanea dentata, 202
Celastrus orbiculatus, 78
Celtis occidentalis, 85
Centaurea cyanus, 7
Cephalanthus occidentalis, 124
Ceratoides lanata, 138
Chamaenerion angustifolium, 117
Chestnut blight, 202
Chronosequence, 29
Clipping, 52, 53, 60, 80
   frequency of, 61–4
   species diversity and, 64
   timing of, 61–4, 65, 67
Clonal growth, 55, 57
   patterns of, 51
   ramets and, 52–3
   ramet integration and, 54, 57
Cohort senescence, 55

Competition, 3, 20, 26, 46, 60, 72, 77, 109, 128, 135, 149
Comptonia peregrina, 80
Controlled colonization, 11, 14–5, 17, 44–6, 77, 117, 124
Controlled species performance, 11, 18, 117, 124, 140
Cooperative research, 20–1, 32
Corridors, 190, 200
  animal migration, 191
  line, 190
  stream, 190
  strip, 190
Cover crop, 135, 140
Crataegus monogyna, 114
Cumulative effects analysis, 212, 215

Danthonia spicata, 80
Deforestation, 7, 8
Demonstrated research, 22
Deschampsia flexuosa, 83
Descriptive research, 20
Dichrostachys cinerea, 87
Digitaria pentzii, 89
Dillwynia peduncularis, 108
Disturbance
  anthropogenic, 7–8, 146, 179–80, 185, 193, 195–8, 217
  designed, 11–3, 15, 17, 47
  natural, 13, 28
  patch, 181
  schemes, 181–3
  soil, 193
Drawdown
  duration of, 120–4
  shoreline succession after, 124–5
  species responses to, 117
  timing of, 118

Edge effect, 185, 186, 187–8, 198
  birds, 156–7
Edge habitat, 8, 156–7, 186–7
Edge species, 186–7
Empetrum hermaphroditum, 124
Eragrostis lehmanniana, 87
Eriophorum vaginatum, 91, 169
Erucastrum gallicum, 55
Eutrophication, 112–3, 114, 193
Exclosure, 165–6
Expert system, 209–11, 216

Fagus grandifolia, 76, 190
Fertility reduction, 113
Fertilization
  of disturbed soils, 109–10
  duration of, 109
  and grass dominance, 108, 109, 110–11
  of grassland, 110–11
  of heathlands, 106–8
  nitrogen, 104, 106–7, 108, 109, 111, 115, 117, 125, 132, 198
  of old fields, 104–6
  timing of, 111
Festuca asperula, 169
Festuca elatior, 80
Festuca ovina, 61, 108
Festuca rubra, 110
Fire
  firelines and, 96–7
  frequency of, 85, 89, 91–2, 95
  fuel and, 91, 92, 95
  ignition patterns and, 97
  interaction between grazing and, 117, 167
  plant adaptations to, 84–5
  resistance to, 85, 86, 92, 101
  timing of, 88–9
  weather and, 96

wildlife habitat and, 95, 159, 162
woody plants and, 85–7
Fluvial processes, 197
Forest
  clear cutting of, 67, 161, 163, 191, 204
  conversion of, 7
  cutting regimens of, 66–7, 153, 161, 178, 202
  as deer habitat, 160–1
  as marten habitat, 163
  as moose habitat, 162
  old growth, 162
  regeneration of 67, 153
  resilience of, 72–3
  secondary succession of, 27–8, 43, 67, 175–6
  as spotted owl habitat, 163
  structural elements of, 155–6, 215
Fruit display, 176

Gaylussacia baccata, 78
genet, 37
Geothlypis trichas, 153
girdling, 72
Grassland
  cattle grazing of, 171–4
  chalk, 61
  clipping regimens of, 60–4
  elephant browsing of, 174–6
  eutophication of, 196–7
  fertilization of, 110–11, 198
  grazing systems of, 172–4
  irrigation of, 114–16
  prescribed burning of, 86–7, 88–9
  rabbit grazing of, 166–7
  sheep grazing of, 167–9
  woody plant invasion of, 85–7, 169, 172, 175, 193
Grass competition, 109, 166

Grazing, 60, 163–4
  by cattle, 171–14
  pressure, 169, 172
  by rabbits, 166–7
  resistance, 169
  by sheep, 167–70
  systems, 169, 178
  value, 98–9, 207–8

Habitat
  for animals, 152–63, 207–8, 212–15
  diversity, 162, 183, 193
  fragmentation, 156, 157, 163, 186, 190, 215
  island, 191
Heathland, 7, 44
  Calluna cycle in, 89
  Calluna regeneration in, 65
  fertilization of, 45–6, 106–8, 159
  grass invasion of, 108
  herbicides in, 83
  irrigation and fertilization of, 116–17
  prescribed burning of, 89–91, 159
  seedling establishment in, 65
  selective cutting of, 64–5
  tree invasion of, 64, 83, 91, 172
Herbicide
  application methods of, 74–6, 77, 80
  drift, 78
  in forests, 75–6
  in heathland, 83
  plant responses to, 75–6
  in rangelands, 83
  in rights-of-way, 77–81
  selectivity of, 73, 100, 140
  and target species, 73–4, 78
Homoranthus virgatus, 108

Hymenoclea salsola, 196
Hyperthalia dissoluta, 89

Ilex glabra, 92
Individualistic response, 73–4, 84–5, 88, 104, 117–18, 155, 167
Initial floristics, 3, 26, 142
Introduced species, 192–5
  controlling, 195, 210–11
Inundation tolerance, 121–3
Invasive species, 192–5
Island biogeography, 183

Juncus squarrosus, 167
Juniperus deppeana, 66, 99
Juniperus virginiana, 65, 85

Kalmia latifolia, 78

Lagopus lagopus, 89, 158
Landscape elements, 180–1, 186, 190
Larrea tridentata, 196
Legumes, 109, 132, 140
Lepus cuniculus, 166
Lespedeza stipulacea, 140
Lespedeza striata, 140
Leucopogon ericoides, 108
Liriodendron tulipifera, 8, 73
Litter, 44, 72, 84
Lolium multiflorum, 140
Lonicera maackii, 53, 193
Loxodonta africana, 174–6

Macrofossils, 31–2
Management goals, 32, 59, 218
  for heathland, 91, 108, 159
  for nature reserves, 180, 212
  for rangeland, 95
  for revegetation, 127–8, 132, 144
Management unit patches, 181, 183, 198, 212–3
  boundary form of, 187–8
  connectedness of, 189–91
  interactions among, 190
  size of, 183–9
Medicago sativa, 146
Mark Twain National Forest, 212
Martes americana, 163
Minespoil
  competitive interactions on, 135, 140, 145
  exotic grasses and legumes on, 132
  fertilization of, 109–10, 128
  grass dominance on, 109, 135
  herbicides in reforestation of, 140
  irrigation of, 116
  native species on, 133–41
  reforestation of, 128, 140–1, 145–6
  species diversity on, 110, 132, 144, 146
  succession on, 146–8
  topsoils on, 141–6
  unmanaged succession on, 146–8
Models, 198, 201–9, 210
  DYNAST, 208, 215
  fire, 205–6
  FORET, 202
  gap, 202
  KIAMBRAM, 202–4
  landscape, 207
  limitations of, 208–9
  PATRECT, 215
Mortality, 26, 121
Mowing, 60
Mulches, 129–30

Nardus stricta, 167
National Forest Management Act, 212
Native species, 133–8, 145–6
Natural disturbance patches, 188–9
Nature reserves, 179
  management of, 180
Nitrogen fixation, 132
Nutrient accumulation, 112, 114
Nutrient exhaustion, 111–14, 126
  techniques of, 113–14
Nutrient use, 103–4

Odocoileus virginianus, 160
Off-road vehicles, 195–6
Old fields, 26–7
  bird populations of, 153
  fertilization of, 104–6, 126
  reversed succession of, 126
Old growth, 156
Opuntia arbuscula, 99
Opuntia fulgida, 99
Oryzopsis hymenoides, 109

Paleoecology, 31–2
Patch
  disturbance, 181
  processes, 188–9
Permanent plots, 23–9, 33
  boundaries of, 25
  layout of, 25
  measurements in, 26, 35–6
Phasianus colchicus, 151, 158
Picea glauca, 46
Pine forests
  prescribed burning of, 91–5
Pinus echinata, 91
Pinus sylvestris, 91
Pinus taeda, 27, 91
Plantago lanceolata, 7
Plantago major, 7

Plant birth, 36
Plant death, 36, 55–6
Plant longevity, 55
Poa pratensis, 88, 115
Poa sandbergii, 83
Pogonarthria squarrosa, 89
Population structure, 29, 36, 38–9, 53
Populus balsamifera, 46
Populus tremuloides, 50, 71–2
Populus trichocarpa, 29
Potentilla erecta, 167
Prescribed burning, 84, 87, 91, 95–7, 100
Propagule bank, 15, 54, 127, 141, 142, 148–9
Prunus serotina, 69
Pteridium aquilinum, 64, 91

Quercus laevis, 75

Ramets, 37, 54
  managing populations of, 65
Rangeland
  controlling tree invasion in, 65–6, 81–3, 193
  grazing systems in, 169, 171–4
  grazing value of, 65, 98–9, 207–8
  rabbits in, 167
Reforestation, 139–40
Regeneration cutting, 67
Remnant patch, 181, 182
Retrogression, 164, 167
Revegetation, 127, 146–8
Rhus, 20
Rhus coppalina, 85
Rhus trilobata, 99
Richmondena cardinalis, 153
Rights-of-way, 77–81
Robinia pseudoacacia, 140, 147
Root kill, 74, 75

Root suckers, 71–2
Rules system, 210

Safe site, 43, 44, 64, 65, 67, 129, 171
Sagittaria latifolia, 118
Salix alaxensis, 46
Scrublands
  grazing value of, 65
  herbicides in, 83
  selective cutting of, 65–6
  tree invasion of, 65–6, 81–3, 193
Secondary succession, 72–3, 181
  and birds, 153–6
Seed bank, 3, 15, 54, 127, 141, 204
  size, 40–3, 57
Seed dispersal, 20, 40, 67, 86, 142, 147, 176–7, 190, 204
Seed dormancy, 43–4, 84, 139, 204
Seed germination, 26, 44, 84, 129
Seeding
  broadcast, 128, 139
  drill, 128, 139
  hydro-, 128–9
  timing of, 140
Seedling establishment, 28, 38, 44–7, 67, 117, 118, 124, 127, 129, 130, 145, 169, 204
Seed mixtures, 130–41
Seeds, 40
Seed production, 47–8, 204
  age-dependent, 48
  mast fruiting and, 48
  size-dependence of, 48
Seed trees, 67
Selective cutting, 27, 67
  in forests, 69–70
  in heathland, 64
  in scrublands, 65–6

Serotiny, 84, 205
Sewage sludge, 105–6
Shelterwood cutting, 67
Shoreline colonization, 124
Smilax rotundifolia, 78
Snow compaction, 196
Soil compaction, 196
Soil erosion, 129, 132, 148
Solidago canadensis, 37, 54, 104–5
Sorghastrum nutans, 88, 115
Sparganium eurycarpum, 118
Species diversity, 26, 65, 132, 135, 153, 167, 169, 172, 175, 182, 183, 188, 193, 196, 212, 215, 218
Sporobolus asper, 115
Sporobolus pyramidalis, 175
Sporobolus robustus, 175
Stable communities, 77–81
Stipa viridula, 109
Stored carbohydrate, 54, 60
Strip corridor, 190
Strip cutting
  and deer use, 161
Strix occidentalis, 163
Stump sprouts, 67–73, 76, 77, 80
Succession
  accelerated, 138–46, 149, 163, 172, 176–7, 196
  and animals, 151–63
  inhibited, 5, 77–81, 85–7
  initial floristics of, 3, 26
  facilitated, 5
  management of, 9–10, 12, 17, 59, 127, 180, 181–3
  relay floristics in, 3
  reversed, 64, 106, 215–6
  theory of, 2–6
  tolerance during, 5, 47
  and plant vital attributes, 5
Surface Mining Control and Reclamation Act, 127

Sus scrofa, 193
Swamp forest, 121–4

Tacking agents, 130
Target species, 73
Taxodium distichum, 124
Terminalia sericea, 87
Themeda australis, 108
Top kill, 74, 75
Topsoil, 141–16, 149
Tree die-back, 55, 198
Tree invasion, 77, 78, 83, 85–7, 121, 124, 147–8, 174–6, 193, 196
Tsuga heterophylla, 29

Ulmus americana, 85

Vaccinium angustifolium, 80
Vaccinium vitis-idaea, 117, 124

Vegetative reproduction, 60, 64, 67–73, 74, 100, 196, 204
  by basal buds, 50, 53, 64, 84, 95
  by fragmentation, 49, 99
  by layering, 50
  by suckering, 50
  by rhizomes, 49
  by tillering, 49, 61, 89, 135
Vegetation
  management, 9–10, 59
  structural components of, 153, 155–6, 177, 178
Vital attributes, 5–6, 204–6

Water level, 114
  manipulation, 117–25, 126, 196–7
Wetlands, 118–21

Zerna erecta, 61